Heritage-Making in Hong Kong Through Culture and Religion

Plate 1 Abandoned house, village of Yim Tin Tsai, Saikung District, Hong Kong, with a statue of Saint Joseph Freinademetz that is installed in the village church

Trevor Sofield · Lawal Mohammed Marafa ·
Fung Mei Sarah Li · Kwo Fung Shek

Heritage-Making
in Hong Kong
Through Culture
and Religion

palgrave
macmillan

Trevor Sofield
Geography and Resource
Management
Chinese University of Hong Kong,
Sha Tin, New Territories
Hong Kong

Lawal Mohammed Marafa
Geography and Resource
Management
Chinese University of Hong Kong,
Sha Tin, New Territories
Hong Kong

Fung Mei Sarah Li
North District, New Territories
Hong Kong

Kwo Fung Shek
Geography and Resource
Management
Chinese University of Hong Kong,
Sha Tin District, New Territories
Hong Kong

ISBN 978-981-97-4338-4   ISBN 978-981-97-4339-1   (eBook)
https://doi.org/10.1007/978-981-97-4339-1

Cover credit: © John Rawsterne/patternhead.com

This Palgrave Macmillan imprint is published by the registered company Springer Nature
Singapore Pte Ltd.
The registered company address is: 152 Beach Road, #21-01/04 Gateway East, Singapore
189721, Singapore

If disposing of this product, please recycle the paper.

CONTENTS

LIST OF FIGURES

CHAPTER 1

Introduction

Abstract To set the context for our research into the rejuvenation of an abandoned Hakka village in Hong Kong, we review a wide set of literature covering heritage-making and religion/religious tourism, cultural tourism, traditional Chinese and Hakka values, spirituality and pilgrimage, sacralization and secularization.

Keywords Heritage-making · Heritagization · Intangible heritage · Religious tourism · Cultural tourism · Pilgrimage · Spirituality · Catholicism · Sacralization · Secularization

To set the context for our research into the rejuvenation of an abandoned Hakka village in Hong Kong, we review a wide set of literature covering heritage-making and religion/religious tourism, cultural tourism, traditional Chinese and Hakka values, spirituality and pilgrimage, sacralization and secularization. It is axiomatic that a fundamental characteristic of community-based tourism is the place where its members reside, since it is the combination of place-plus-people that constitutes the foundation for this type of tourism. Yet amidst the still-extant ruins of Yim Tin Tsai a unique form of sustainable community-based tourism that combines natural and built heritage, Catholic Christian pilgrimage, Hakka (Chinese) culture, traditional cottage-industry salt-making, and in the

1

T. Sofield et al., *Heritage-Making in Hong Kong Through Culture and Religion*, https://doi.org/10.1007/978-981-97-4339-1_1

past six years a Government-sponsored arts festival, has arisen from an abandoned, almost-forgotten place.

The island's resurrection involves community families who are not only Hong Kong residents but includes its internationally-dispersed diaspora. The attachment of the original inhabitants to place was waning steadily until 21 years ago when, through a mode of co-creation involving the former residents and the broader Catholic community of Hong Kong, allied to a pivotal (but unknowing) input from the Vatican in Rome in 2003 and a supporting role played by UNESCO, Bangkok, in 2005, the community was revitalized through projects focused on restoration. These included *inter alia* major renovation of its semi-derelict 1889 church, transformation of the desolate primary school into a Hakka Exhibition Centre and folk museum, and subsequently the restoration of community salt pans (abandoned around 1910) which had been overgrown with a mangrove forest. For the first 13–14 years heritagization was not commoditized because of the centrality of motivation derived from Catholicism which focused almost exclusively on pilgrimage. The island's pilgrimage development was overseen by a Joint Organizing Committee composed of the Village Council and the Catholic Diocese of Hong Kong, which established a non-profit non-Governmental organization (NGO) in 2013 called "Salt and Light" to manage visitation. There was no promotion of Yim Tin Tsai as a general sightseeing place, there was no entry fee, and management and maintenance costs were dependent upon donations channelled into a trust fund. It was not until the Tourism Commission of Hong Kong, seeking to broaden the attractions and appeal of Hong Kong, saw the potential of Yim Tin Tsai in 2017 and opened discussions with the Joint Organizing Committee to expand visitation beyond pilgrimage. Until that point, entanglement was largely irrelevant as an analytical tool for probing what was happening on Yim Tim Tsai. Currently, while Catholicism retains its primacy in many respects and a degree of segregation (separating pilgrims from non-pilgrims) occurs, a majority on the Village Council has demonstrated a willingness to expand secular activities organized by the Hong Kong Tourism Commission, partly because of generous Government funding assistance. Propelled by the impetus of the Tourism Commission's intervention, collaborative tourism management partnerships involving the public sector, the island community, the Catholic Church of Hong Kong and NGOs have emerged. "Salt & Light", originally established to conduct only pilgrimage management and tours of sacralized sites, now has oversight of all tourism to the island which includes commoditized forms of entrance-fee heritage tourism to the salt pans and the Hakka Exhibition Centre. Where previously "Salt & Light" relied heavily upon volunteers it now has less docents as it juggles pilgrimage visitation with commercial activities. In temporal terms our investigation thus finds itself *in medias res* as the situation on the island continues to evolve. While we foreground 'heritage-making' as a verb, and 'heritage' as a dynamic process (following Harvey, 2001) in our study of Yim Tin Tsai, we stress that many of the outcome(s) involve ongoing adaptation and changes to both physical properties and interpretation carried out from and at the grassroots local level. This is in contrast to heritage on a national scale that often seeks to appropriate cultural capital as a finished product in the context of competing political claims associated with power. As Hui et al. (2018, n.p.) argue, *heritage becomes the manifest material and symbolic anchor for culture, and one must have a "heritage" as one must have a nose and two ears (to borrow Gellner's simile) if one is to be recognized and recognizable in the international, national and sub-national arenas.*

This case study of Yim Tin Tsai encapsulates three well-defined areas of current research in the heritage literature: sacralization/secularization, religious heritage and the question of renewal often associated with a quest for spirituality and pilgrimage (de Jong & Mapril, 2023). These are clearly manifested in Yim Tin Tsai. When the Joint Organizing Committee seized upon the heritagization of its Catholic legacy as the vehicle to bring life back to the island, especially with its adoption of pilgrimage (consistently described throughout centuries as a journey of renewal: Di Giovine & Garcia-Fuentes, 2016; Di Giovine & Choe, 2020) and the naming and design of its nature walk as 'The Catholic Nature Trail of Reconciliation' (Chan, Chi-Ming, 2012), it launched spiritual renewal, so pivotal to religious and secular ontologies (De Jong & Mapril, 2023) with physical renewal (the restoration of the church) at the intersection of sacralization and secularization.

The deep intricacies within these topics inspired Olsen & Timothy to describe their 2021 *Routledge Handbook of Religious and Spiritual Tourism* also as an exploration into the intersections of religion, spirituality and tourism. With 33 individual contributions their edited volume is ambitious in its scope as it delves into developing issues and trends arising from increasing research in these areas. They note that definitions of religion and spirituality are problematic insomuch as even in theological circles 'religion' has numerous definitions and that 'spirituality' is even more elusive. They identify four main reasons for the lack of specificity. The first is that while religion is found in all cultures, its structures and meanings differ widely. For example, Olsen and Timothy (2021), citing Mitsutoshi Horii, 2018 who researched the meaning of religion as a social category within the sociological context of contemporary Japan, aver that many western societies differentiate between religious and secular spheres, but in other societies there is no such distinction since religion is innately connected to virtually everything, from politics and governance to culture, heritage, health and the economy, defining societal norms and values that set attitudes and guide behaviour. Spirituality is even more difficult to grasp definitively, in part because it is generally perceived as the quest by an individual for meaning and is often pursued and found outside any religion (e.g. Maslow's self-actualization, self-transcendence and spirituality: Koltko-Rivera, 2006), yet this quest is also found in institutionalized religion (Olsen & Timothy, 2021, p. 5). Second, the two terms become more entwined when considered in the context of 'religious tourism' where the supply side and the demand side take variant views on segmentation for marketing purposes. A third difficulty in trying to distinguish between religious tourism and spiritual tourism is because some post-modern academics (see Collins-Kreiner, 2020, for example) argue that "definitive definitions are too constraining", so they create 'stipulative' definitions that are specific to the situations they are researching (Olsen & Timothy, 2021, p. 7). This thus creates a multiplicity of variant definitions. A fourth complication arises from the fact that the religious tourism market has become increasingly fragmented into other separate but over-lapping niche markets: religious tourism, pilgrimage or faith tourism, spiritual tourism and New Age tourism (Olsen & Timothy, 2021, p. 8). They include both supply and demand side perceptions, including built heritage sites, ceremonies and other intangible heritage activities, and motivations for undertaking travel to certain destinations. The spiritual tourism market is closely related to pilgrimage and faith tourism, but as noted it also extends outside formalized religion and some argue that it is the central feature of wellness tourism and New Age tourism. In short, there has been a proliferation of new forms of religion and spirituality, all of which carry the potential for commodification (Olsen, 2019). In the context of Yim Tin Tsai's pilgrimage tourism, we have not pursued spirituality as a separate-but-interlinked phenomenon with religious tourism although our immersion in the island setting over a number of years has exposed us to many instances where individuals have undergone uplifting experiences that they have expressed in terms of spirituality (although not as an intense transcendental experience), and it offers a new area for future research.

Olsen and Timothy (2021) have divided the chapters of their volume into four sections. The first section deals with definitions, concepts and theories of religious and spiritual tourism; a series of six chapters in Sect. 2 explores religious places and spaces that lean towards a supply side analysis that broadens spiritual spatiality to include national parks as treasured sacred places; Sect. 3 by contrast takes a demand side approach to examine motivation, experiences and performance across five diverse chapters; and in Sect. 4, management issues related to religious and spiritual tourism are covered in depth. All of these sections to varying degrees have relevance for our research into the physical and spiritual renewal of Yim Tin Tsai as a pilgrimage destination, a religious space and place, a diverse experiential environment for tourists and now a destination cross-cut with new forms of heritagization that attract secular tourists; and as we have proceeded with our analysis of Yim Tim Tsai in the chapters that follow below, we have drawn on relevant contributors to this volume to enhance our own understandings.

In this context, many local participants and observers tend to use the binary 'sacred/secular' construct to approach the dynamics of Yim Tin Tsai and we tend to follow that distinction in our discussions on the situation as a useful shorthand. We do suggest, however, in agreement with Balkenhol et al. (2020, p. 1) that *"ways in which sacrality and secularity mutually inform, enforce and spill over into each other"* where their *boundaries overlap, are claimed, contested, reclaimed and re-contested in new ways,* tend to diminish *"categories of religion and secular as neutral descriptive terms"*. More than 40 years ago, anthropologists like Turner (1974a, 1947b) and Graburn (1977) were describing tourism as 'the sacred journey', equating non-religious tourism to pilgrimage in part because of uplifting or transformative experiences that occur outside the institutions of formal religions. Griffin and Raj (2017, p. i) note that: *"Opinions are divided on whether visiting a hero's grave"* (e.g. the mausoleum of Mao Zedong lying in state in Tien An Men Square, Beijing, or Ho Chi Minh in Hanoi); *"a site of environmental/human tragedy" (e.g. Pompeii in Italy"*/Hiroshima in Japan*); "a battlefield site" (*e.g. the Battle of the Red Cliffs, Yangtze River near Wuhan in Hubei Province China 208–209 A.D., regarded as the most famous battle of ancient China*); "or an ancestral home (diaspora returning to their ethnic place of origin"*, such as the Hakka returning to Yim Tin Tsai), *"can be considered as pilgrimage"*. Joining the long queues that assemble daily in Beijing and Hanoi and observing the reverence with which local visitors pay their respects to the embalmed bodies of Mao Zedong and Ho Chi Minh leaves no doubt that for them it is a deeply moving and religious experience. *Increasingly, travel which purposely or inadvertently includes a meaningful, transformative experience beyond the norm that impacts an individual's belief system is being recognized as secular pilgrimage* (Griffin & Raj, 2017, p. iii); and tourism that imbues virtually any travel experience with spirituality may also be considered akin to religious tourism (Olsen & Timothy, 2021). Melanie Smith (2021, web page) argues cogently that religion, spirituality and wellness tourism have much in common:

> Spirituality and wellness are inextricably connected. Wellness can be defined as the path to achieving well-being, a path that includes physical, mental, and spiritual health, self-responsibility, social harmony, environmental sensitivity, intellectual development, emotional well-being, and occupational satisfaction, whereas spirituality includes experiencing oneness with nature and beauty and a sense of connectedness with self, others and a higher power or larger reality, concern for and commitment to something greater than self …. Travel has been viewed as a way of enhancing and transforming individual self-development and even changing world views.

With this in mind, while we broadly differentiate between what has been described as sacred and secular heritage and their accompanying forms of tourism, we emphasize that religion/religious sites/ ceremonies and events, whether sacralized or secularized, are cultural heritage and do not exist in some separate silo. In China, religions and popular religious practices, such as processions of local deities, ancestor worship, temple festivals, exorcisms, divination and spirit mediumship, once condemned as feudal superstition by the Communist Government, were embraced as valued culture in 2004. They lost their stigma and were re-branded as 'intangible cultural heritage' when China became a signatory to the UNESCO Convention for the Safeguarding of Intangible Cultural Heritage (UNESCO, 2003) and introduced its intangible cultural heritage inventory system (Zheng, 2023). An interesting affirmation of religion as cultural heritage found expression in Italy where the legislatively-ordained display of Catholic crucifixes on classroom walls was challenged in the European Court of Human Rights in 2016. The final ruling of the Court found that the presence of crucifixes in public schools did not violate students' rights to freedom from religious indoctrination *based on an interpretation of the crucifix as a passive symbol of Italian cultural heritage* (Astor et al., 2017, p. 4). As a 2015 Report adopted by the European Parliament stated: *Religious heritage constitutes an intangible part of cultural heritage … Historical religious heritage, including architecture and music, must be preserved for its cultural value, regardless of its religious origin.*

REFERENCES

Astor, A., Burchardt, M., & Griera, M. (2017). The Politics of Religious Heritage: Framing Claims to Religion as Culture in Spain. *Journal for the Scientific Study of Religion, 56*(1), 126–142. https://doi.org/10.1111/jssr.12321

Balkenhol, M., van den Hemel, E., & Stengs, I. (2020). Introduction: Emotional Entanglements of Sacrality and Secularity—Engaging the Paradox. In M. Balkenhol, E. van den Hemel, & I. Stengs (Eds.), *The Secular Sacred*. Palgrave Macmillan.

Collins Kreiner, N. (2020). Religion and Tourism: A Diverse and Fragmented Field in Need of a Holistic Agenda. *Annals of Tourism Research, 82*(2). https://doi.org/10.1016/j.annals.2020.102892

de Jong, F., & Mapril, J. (Eds.). (2023). *The Future of Religious Heritage: Entangled Temporalities of the Sacred and the Secular*. Routledge.

Di Giovine, M. A., & Garcia-Fuentes, J. M. (2016). Sites of Pilgrimage, Sites of Heritage: An Exploratory Introduction. *International Journal of Tourism Anthropology, 5*(1–2), 1–23.

Di Giovine, M. A., & Choe, J. A. (2020). Geographies of Religion and Spirituality: Pilgrimage Beyond the "Officially" Sacred. *Tourism Geographies, 21*(3), 361–383. https://doi.org/10.1080/14616688.2019.1625072

Graburn, N. H. H. (1977). Tourism: The Sacred Journey. In V. Smith (Ed.), *Hosts and Guests: The Anthropology of Tourism* (pp. 17–31). University of Philadelphia Press.

Griffin, K., & Raj, R. (2017). The Importance of Religious Tourism and Pilgrimage: Reflecting on Definitions, Motives and Data. *International Journal of Religious Tourism and Pilgrimage, 5*(3). https://doi.org/10.21427/D7242Z

Harvey, D. C. (2001). Heritage Pasts and Heritage Presents: Temporality, Meaning and the Scope of Heritage Studies. *International Journal of Heritage Studies: IJHS, 7*(4), 319–338. https://doi.org/10.1080/13581650120105534

Hui, Y.-F., Hsiao, H.-H. M., & Peycam, P. (2018). Introduction: Finding the Grain of Heritage Politics. In H. Hsiao, Y. Hui, & P. Peycam (Eds.), *Citizens, Civil Society and Heritage-Making in Asia* (Ch.1). ISEAS Publishing.

Koltko-Rivera, M. E. (2006). Rediscovering the Later Version of Maslow's Hierarchy of Needs: Self-Transcendence and Opportunities for Theory, Research, and Unification. *Review of General Psychology, 10*(4), 302–317. https://doi.org/10.1037/1089-2680.10.4.302

Olsen, D. H. (2019). Religion, Spirituality, and Pilgrimage in a Globalizing World. In D. J. Timothy (Ed.), *Handbook of Globalisation and Tourism* (pp. 270–283). Edward Elgar.

Olsen, D. H., & Timothy, D. J. (2021). *The Routledge Handbook of Religious and Spiritual Tourism*. Routledge.

Smith, M. K. (2021). A New Spiritual Marketplace: Comparing New Age and New Religious Movements in an Age of Spiritual and Religious Tourism. In D. Olsen & D. Timothy (Eds.), *The Routledge Handbook of Religious and Spiritual Tourism* (Ch. 6). Routledge.

Turner, V. (1974a). Pilgrimage and Communitas. *Studia Missionalia, 23*, 305–307.

Turner, V. (1974b). *Dramas, Fields, and Metaphors: Symbolic Action in Human Society*. Cornell University Press.

UNESCO. (2003). *Operational Guidelines for the Implementation of the World Heritage Convention*. UNESCO.

Zheng, S. (2023). The Heritagisation of Rituals: Commodification and Transmission. A Case Study of Nianli Festival in Zhanjiang, China. *Études Mongoles & Sibériennes, Centrasiatiques & Tibétaines, 54*. https://journals.openedition.org/emscat/6109?lang=en

CHAPTER 2

The Concept of 'Heri-ligion'

Abstract To explore the intricacies, processes and forces at work in the resurrection of Yim Tin Tsai from an abandoned Hakka Catholic village to a 'new heritage' site encompassing both tangible and intangible religious and secular characteristics, we advance a concept that we have called 'Heri-ligion'. It has multiple facets that combine, interact and are entangled in a variety of ways. These facets include inter alia attachment to place and 'home', religious affiliation and its accompanying value system, syncretism and symbiosis, politics and power, stakeholder analysis, co-creation, socio-anthropological entanglement theory and environmental impacts and sustainability.

Keywords Religious heritage · Cultural heritage · Sacred · Secular · Place · Home · Collective memory · Stakeholder analysis · Co-creation · Syncretism · Symbiosis · Politics · Entanglement

To explore the intricacies, processes and forces at work in the resurrection of Yim Tin Tsai from an abandoned Hakka Catholic village to a 'new heritage' site encompassing both tangible and intangible religious and secular characteristics, we advance a concept that we have called 'Heri-ligion'. It has multiple facets that combine, interact and are entangled in a variety of ways. Our term is derived from an EU

© The Author(s), under exclusive license to Springer Nature Singapore Pte Ltd. 2024
T. Sofield et al., *Heritage-Making in Hong Kong Through Culture and Religion*, https://doi.org/10.1007/978-981-97-4339-1_2

anthropology-oriented research project, *The Heritagization of Religion and the Sacralization of Heritage in Contemporary Europe* (2016–2019), which had a focus on utilizing the construct of 'entanglement' as a key analytical tool (https://anthropology.ku.dk/research/research-projects/completed_projects/hereligion/). The team of European anthropologists gave their project the acronym of 'hereligion' without developing the term as a concept in its own right. Their research endeavours focused on identifying 'problem-areas' wherein heritagization and sacralization were somehow 'entangled',[1] using this term in its anthropological/sociological sense. Through the lens of entanglement and resultant tensions they sought *"to understand the consequences of the heritagization of religious sites, objects and practices which may not have been considered heritage before, and especially the relations between heritage and religious constituencies, and between different disciplines and management regimes; and the potential paradoxes between religious and secular sacralizations and uses"* (Professor Oscar Salemink, leader of the EU project, personal correspondence 2019). Via a series of research projects in the Netherlands, Denmark, UK, Poland and Portugal, a focal point was to appraise the ways in which aspects of what are now considered secular heritage still retain religious and possibly sacred significance for some people and communities and how such sites are managed when confronted with different, often conflicting, demands from their diversely-motivated visitation, on the one hand, and local residents, on the other.

The tensions associated with and often regarded as inherent in entanglement have led to the term acquiring generally negative connotations, as in 'a predicament', 'a mess', 'an imbroglio', 'a compromising relationship or situation', 'entrapment' (Oxford English Dictionary, 2012). In our concept of heri-ligion, however, we seek to balance this negativity by ensuring that other forms of entwining co-involvement may be an outcome, such as in examples of syncretism and symbiosis where, by contrast, relative harmony occurs.

Syncretism is defined as: *the amalgamation/assimilation of several originally discrete traditions, especially in the theology and mythology of*

---

[1] 'Entanglement' is a concept in contemporary anthropology and sociology, the dialectic of dependence and dependency between human-to-humans and humans-to-things. The term "entanglement" seeks to capture the ways in which humans and things entrap each other, and identify the tensions that arise (Hodder, 2023).

religion, thus asserting an underlying unity and allowing for an inclusive approach (Oxford English Dictionary, 2012). Chinese belief systems and philosophies, for example, have a long history of syncretism incorporating elements of ancestral worship and divination from the Shang and Zhou Dynasties (circa 1600–256 BCE), with Confucian Thought (551–479 BCE), Daoism and Buddhism that was introduced into China 2000 years ago. Elements from these mainstream faiths and value systems are intermixed in a wide variety of practices and ways that find expression in contemporary China (Ching, 1993). It is not unusual in religious practice in China today to see Buddhist and Confucian artefacts, statues and dioramas in a Daoist temple and vice versa, nor a Buddhist colleague propitiate an historical figure famous for his Confucian virtues at a Daoist temple (personal observations by co-authors Sofield & Li over 30 years of field trips around China). The constant elevation of 'harmony' during millennia as one of the most important of all Chinese values is considered to be a major contributing factor to the propensity for syncretism that has been present across the centuries:

> Unlike some other cultures, where religious syncretism and even tolerance are viewed with skepticism or condemnation, the Chinese have always had the ability to select the religious practices and teachings that work best for them at the time. … In general religious pluralism simply adds to the many options from which the Chinese can choose on their journey toward a harmonious life. (Foy, 2023, n.p)

In this context, traditionally the religious folklore of the Hakka amalgamated forms of ancestor worship, Buddhism, Daoism and a generalized Confucian worldview, with one distinctively Hakka religious practice that involved the worship of dragon deities, especially the earth dragon god because of their monolithically rural farming lifestyle over centuries where the land was fundamental to their survival (Lin Zuolao, 2022).

Symbiosis is a collective term drawn from ecological science and refers to situations where organisms which inhabit the same space or environment may share or compete for the same resources and interact in five main types of relationships:

(i) *mutualism*, where interaction is beneficial to both organisms;
(ii) *commensalism*, an association between two organisms in which one benefits and the other derives neither benefit nor harm;

(iii) *predation*, where one organism kills and consumes another;
(iv) *parasitism*, a non-mutual relationship between two organisms in which one benefits at the expense of the other (the host); and
(v) *competition*, when two organisms compete for the same limited resource/s.
(vi) mutualism, commensalism and parasitism may also be considered different forms of *co-existence*, where both organisms exist side by side ranging from neutral tolerance or degrees of benefit to one or the other.

Collectively the organisms/parties involved in a symbiotic relationship are called symbionts (Douglas, 1994, 2010; Martin & Schwab, 2012).

These types of interactions may validly be applied to human interactions, e.g. the Oxford English Dictionary (2012) defines symbiosis as: *a mutually beneficial relationship between different people or groups*, and Sagarin (2013) has suggested symbiosis as a business strategy to develop greater efficiencies and synergy between suppliers of products and distribution networks. This does not mean that we ignore conflict inherent to entanglement situations since it is an appropriate tool to analyse some aspects of a given situation. For example, parasitism when extended to consider how the growth of mass tourism may feed off a religious site and/or tradition and/or custom to become a highly commercialized, commoditized secular sight, may be deemed another aspect of entanglement. The pilgrimage to Santiago de Compostela in Spain is one such example, where secular tourists now outnumber pilgrims by about ten to one. The Catholic authorities manage access to the Santiago cathedral and still operate many hostels along the pilgrim's way, the *camino* of St James,[2] but the tourism and travel industry and FITs (free independent travellers) leverage their tours off the religious traditions, connotations and heritage of the *camino* of Santiago, often as 'adventure tourism' in a secularized commodification of a sacred ritual and sacralized sites (Farias et al., 2019; Murray, 2021). In contrast with general sightseeing tourists,

---

[2] The *camino* of St James is the Pilgrims' Path or Holy Way that runs for more than 1700kms along a centuries-old route from Paris to Santiago de Compostela in northwestern Spain, although there are other shorter routes that start, for example, on the French side of the Pyrenees between Bayonne and Perpignan and are between 400 and 800kms long. Other routes start in Portugal and Italy.

religious pilgrimage tourism may often differ significantly from other forms of tourism because the motivation for attending/participating in religious festivals is different and pilgrims will not necessarily be enticed to visit other attractions or seek other activities (e.g. Bauman, 1996; Farias et al., 2019; Santos, 2002).

Heritage formation as defined by Meyer and de Witte (2013, p. 1) will often embrace politics as:

> "... a complicated, contested political–aesthetic process. Which aesthetic practices are involved in profiling cultural forms as heritage? What are the politics of authentication that underpin the selection and framing of particular cultural forms? To which contestations does the sacralization of particular cultural forms—in particular, those derived from the sphere of religion—give rise? Which aesthetics of persuasion are invoked to render heritage sacred for its beholders, be it "the world" (as addressed by UNESCO World Heritage schemes), the "nation" (as addressed by national heritage sites), or smaller constituencies, including minority groups for whom claiming heritage may be designed to give them a voice in the political arena?

Politics and power relations are inevitably present in virtually all heritagization situations but are potentially more visible at national levels where heritage may be invoked to capture and define national identity for the benefit of one political persuasion over another. Globally, all national monuments and museums established and funded by governments will, as a guiding principle, attempt to embrace an authentic national identity which often reflects the ideology of the Government in power. The former Soviet Union, for example, held that museums were entirely secular vehicles for the promotion of Government policy: religion was interpreted in Marxist evolutionary terms as outdated, non-scientific mythology that impeded modernization and progress, and they established a special category of "museums of atheism" to eradicate or marginalize religion (Deschepper, 2018; Hirsch, 2014; Luehrmann, 2015). Interestingly, since the downfall of the Soviet Union, in Russia and the former Soviet republics where there has been a massive surge in and re-engagement with religions in public life (Orthodox and Catholic Christianity, Buddhism, Islam, depending upon locality), heritagization of religion has become a central part of national and ethnic museum exhibitions as these countries seek to throw off the repression of Soviet colonialism and forge their non-Soviet identities in their own self-image

(Bogumil & Lukaszewicz, 2018, Pimenova, 2022). Many churches and other religious buildings throughout the former Soviet Union had been restored and re-purposed as 'museums of atheism' rather than destroyed (as occurred with the Cultural Revolution in China between 1966 and 1976, for example) and they have been re-appropriated by religious institutions, aiding in the rejuvenation of religions throughout the new Russia. In short, heritage-making, and especially the processes involving heritagization of religions, are embedded in cultural politics at multiple scales (Harvey, 2015), but some of the stakeholders at different scales may become involved with different degrees of agency and interests that cut across scales (Hui et al., 2018; Peycam et al., 2020). In Yim Tin Tsai for example, at the international scale there is UNESCO and the Vatican; at the national level there is the Hong Kong Government and the Catholic Diocese of Hong Kong; at the district scale there is the Saikung District Council; and at the local level there is the Village Rural Committee and the grassroots community members.

In Yim Tin Tsai village, localized politics are unavoidable in heritage-making but the role of the Catholic faith in moderating conflict and seeking compromise, in terms of inter-personal conflict or inter-institutional differences, has been a steadying influence. Village governance through an elected Council Representative, as enacted via the Rural Representative Election Legislation (Amendment) Bill 2013, provides a legal basis for the 693 indigenous villages in Hong Kong *inter alia "to deal with all affairs relating to the lawful traditional rights and interests, and the traditional way of life, of those indigenous inhabitants"* (Home Affairs Department, Government of Hong Kong official website, 2023). It is this statute and accompanying legislation that guarantees land ownership and inheritance rights for indigenous villages which has provided the cultural space for the Yim Tin Tsai community in cooperation with the Catholic Diocese of Hong Kong to proceed with its heritagization projects relatively free from central government oversight. National policies and politics, which in many countries would be possibly the single most important determinant factor in charting heritagization, are characterized by a non-interventionist stance in Yim Tin Tsai where even village politics have been a minor factor in its development. Niedźwiedź and Saraiva (2015) in analysing religious heritagization draw a distinction between bottom-up grassroots activities that operate at the local level and top-down Government and intergovernmental agencies that function on a regional, national, trans-national or global level. The religious heritagization on Yim Tin Tsai does not display such a clear dichotomy. The Catholic Church through the Hong Kong Catholic Diocese and indirectly the Vatican in Rome may be classified as top-down actors but the Vicar General of the Hong Kong Catholic Diocese in addition to his bishopric mitre also wears the hat of a member of the local Hakka community, and the restoration of the church and the establishment of the Pilgrim's Trail on Yim Tin Tsai and accompanying sacralization of parts of the village environs required intense and intimate cooperation with the Village Committee whose elected Council representative had been promoting restoration of the village for years. Hence a fundamental necessity of religious heritagization on the island came from the grassroots level. Together the Hakka community and the Catholic Diocese established their Joint Committee and they proceeded with their action plan for religious heritagization without political, policy or financial involvement of the Hong Kong Government. Subsequently 'secular' heritagization that broadened the entire heritagization process on Yim Tin Tsai occurred with some direct Government support, but it is important to note the central role that the Joint Committee and the Hakka community exert over religious and secular tourism to Yim Tin Tsai. We argue that it is important to include consideration of power, politics, policies and governance at different scales in our concept of heri-ligion.

In approaching heri-ligion we emphasize that it is bi-directional, i.e. the flow of agency may move in opposite ways. When a former religious site or event has been deconsecrated and re-defined as heritage it incorporates an explicitly secular gaze predicated on non-transcendent principles—historical, cultural, aesthetic. The reverse side of the coin is that there may be processes whereby non-religious aspects of sites may be sacralized and thus endowed with transcendent characteristics (de Jong & Mapril, 2023). Such aspects may be built or intangible, 'capturing' for example a secular structure and transforming it into an object of religious veneration, or a profane event or ceremony that is similarly sacralized. Such a transformation may not necessarily deny its original form so that it may continue to be viewed as secular heritage by one set of observers/ users and simultaneously as embodying sacred elements by another. As Wang et al., (2021, p. 3) note:

> Interaction between heritage and religion creates "a field of cultural production" where values are established as products of encounter and

exchange (Bourdieu, 1993). This interaction creates both overlap and conflict between religious practices of continuity and heritage preservation.

Meyer and de Witte (2013, p. 1) describe how the sacred when 'profanized' and reframed as heritage may be accepted, contested or rejected; and conversely how profane/secular heritage forms when sacralized '*may make them appear powerful, authentic or even incontestable*. In other words, this duality of purpose may lead to tensions between the two differing perceptions as per entanglement, or they may be harmoniously accommodated in an active symbiotic relationship such as mutualism or in passive tolerance such as commensalism.

Religious tourism—pilgrimage in particular—shares aspects in common with heritage tourism, starting with the fact that many travellers will refer to their 'secular' destination as 'a pilgrimage'. According to Di Giovine (2012, p. 117):

> "A pilgrimage is a ritual journal from the quotidian realm of profane society to a sacred center, a passion-laden, hyper-meaningful voyage both inwardly and outwardly, which is often steeped in symbols and symbolic actions.

Catholic pilgrims to Yim Tin Tsai will be able to access a shared identity as adherents of the faith espoused by Saint Joseph Freinademetz and participate fully in the religious ceremonies and other acts of worship constructed around his persona. For those of its former Hakka residents who may no longer be Catholic, and younger generations who were not born and brought up there, similar highly-charged feelings may nevertheless be generated by the sight/sites of their village-based childhoods and/or the ruins of their ancestors' homes. Both may equally use visitation to Yim Tin Tsai to embrace their common identity. They imbue the sight/site with "*a hyper-meaningful quality*" (Di Giovine & Garcia-Fuentes, 2016, p. 3) that 'sacralizes' the site and associated experience(s) in what Turner (1974a, 1974b), Graburn (1977, 2001), Di Giovine (2012, 2021) and others have described as secular pilgrimage, transforming a natural or built heritage site from the mundane, everyday profane to an elevated 'sacred' status.

With reference to secular tourists to the island, their activities may lack the elements of spiritual worship and participation in religious rites that are integral to a religious creed framing a pilgrim's visitation to a sacred site, but the secular tourist may also undergo a spiritual experience when deep meaning and significance is attached to the object of their gaze (Di Giovine, 2021; Di Giovine & Choe, 2019; Olsen & Timothy, 2021). We have experienced this ourselves, for example, when visiting World Heritage-listed sites such as the rainforests of the Virunga Volcanoes National Park in Rwanda to view the wild, free-roaming silver-backed gorillas, or kayaking among whales, seals and penguins amidst the icebergs of Antarctica, or touring the 1000-year-old Khmer ruins of Angkor Wat in Cambodia. Religion and heritage may become entangled with differing directions and objectives when a place is perceived as both a religious pilgrimage site and a secular tourist site as with Angkor (Sofield, 2013), but may equally find areas of reciprocal acceptance. In Yim Tin Tsai, religious pilgrimage and secular heritage visitation at times mutually inform each other as aspects of their attributes overlap to reveal shared phenomenological and ontological characteristics (e.g. when they visit the Catholic cemetery, where most visitors will show respect regardless of religious or atheist affiliation).

For non-Yim Tin Tsai visitors the island's heritage may still resonate with deep meaning. For non-Christian, non-Hakka outsiders, their response to the heritage of Yim Tin Ysai will be coloured by their differing worldviews, so that inevitably their interaction with the site and interpretation of their experience(s) will diverge. But even then, there will be aspects that exhibit commonalities between differing sets of visitors in a symbiotic relationship where a degree of mutual understanding reduces the tensions that are often present when the same site is viewed as both a pilgrimage site and a secular heritage site. And this, of course, serves to emphasize that there is not a dichotomy between religion and heritage, but that indeed religion is one of the purest manifestations of heritage: the places, spaces, architectural forms, rituals, ceremonies, scriptures, music, traditions, prayers and beliefs are all part of cultural heritage (Timothy, 2011).

Our concept of heri-ligion builds on the approach adopted by the EU research teams, and while entanglement remains a significant component in our conversion of the amended acronym to a concept, we extend it to include additional processes:

i. Ways in which religious cultural heritage, both built and intangible, may contribute to defining identities and communities in times of change;

ii. The complexities of place attachment and motivations underlying a community's drive to define and/or re-discover and re-establish its roots ('home') where religion/religious traditions are of foundational importance, that include;

iii. Specific, pre-existing characteristics of a community which may be perceived as major determinants in directing the dynamic pathways of change (ethnicity, cultural value system, geography and so forth);

iv. Collective imagining and memory reconstruction, and nostalgia;

v. The role of co-creation in bringing different parties/agencies to participate in the sacralization or secularization of that place or activity; hence

vi. Incorporation of stakeholder theory and stakeholder dynamics into the concept of heri-ligion;

vii. The interrelationships between power, politics, policies and governance as fundamental, particularly when heritagization takes place at/on a national scale since heritage formation is often a complicated, contested political–aesthetic process, especially when the sacralization of cultural forms encompasses those derived from the sphere of religion;

viii. Recognition that there may often be a bi-directional process, i.e. whereby some secular sites and/or ceremonies may be sacralized by those stakeholders for whom religion is a major driver, and by contrast where some sacred sites and ceremonies may be secularized and commoditized by those stakeholders for whom conservation and/or the construction of tourist attractions are prioritized. We call this 'the sacred-secular gaze' to emphasize the interconnecting processes at work;

ix Application of the components of syncretism and symbiosis to strengthen the analysis of the unfolding dynamics of a given situation; and

x. Employing *inter alia* entanglement theory to decipher the complexity and diversity of the many processes often concurrently at work, asserting a due place for functionalism—the relationship and interdependency between all social groups, big and small.

xi Environmental impacts of religious heritagization: sustainability and regeneration (Ch 13 refers).

Not all of these factors may be present together in any specific situation, but we have compiled a comprehensive set that in our view provides an enhanced capacity to explore and unpack the many processes at work where sacralization and secularization (as differing streams of the assemblage of cultural heritage) coincide.

As a pioneer application of our concept of heri-ligion, we examine the case of Yim Tin Tsai, a 370-year-old Hakka village on an island in Hong Kong that requires the multifaceted approach of all eleven factors to delve into and reach some understanding of its complexity.

The island was completely abandoned in the 1990s but is currently undergoing a rejuvenation initiated in the first instance by religiously-oriented stakeholders and now joined by the secular Hong Kong Tourism Commission. In our analysis, we see how relationships emerge between heritage and religious constituencies and a diversity of stakeholders. We observe how entanglement as a pertinent factor really arose only after the intervention of the Tourism Commission which introduced commodification and secularization of heritage that had originally been largely absent. This intervention has resulted in a re-ordering of the power relationships between stakeholders, with the transmutation of the island from almost exclusively a place of pilgrimage into a sightseeing destination that is overtaking its original Christian-determined alignment.

2.1 METHODOLOGY

A mix of methods has been utilized to document the metamorphosis of Yim Tin Tsai from a derelict place to a dynamic pilgrimage and tourist site with the Hakka Catholic community at its forefront. Critical Realist Philosophy underpins this broad approach: it encompasses a range of interdisciplinary perspectives to facilitate the creation of new theoretical models that draw information from different academic silos which often have a narrow confined focus, and our concept of heri-ligion emerged from this modus operandi. In determining how to undertake an analysis of Yim Tin Tsai, the varied academic backgrounds of all four authors influenced the multi-disciplinary approach that was adopted: qualifications include undergraduate and postgraduate degrees in social anthropology, environmental science, forestry, education, geography, primatology, sustainable tourism management, leisure and recreation, convention and events management, development theory, and international political science and economics. These qualifications have been accompanied by numerous research endeavours in many countries, which include more than 50 published papers on China and Hong Kong. With this expertise available to inform our research it was inevitable that Critical Realist Philosophy became the aegis which allowed interdisciplinary contributions to guide our fieldwork. Critical Realist Philosophy is described by Thouki (2022, p. 1038) as an approach that *"encourages researchers to engage with a broad body of literature without any strict inclusion or exclusion criteria, aiming to combine perspectives and create new theoretical models"* (see also Edgley et al., 2016; Torraco, 2016), and out of which emerged our concept of 'heri-ligion'. It may be considered a form of constructivist grounded theory, a research method that focuses on generating new theories through inductive analysis of data gathered from participants (Charmaz, 2006, 2014). In this context, interviews may capture a certain reality but are better regarded as *"emergent interactions through a mutual exploration of the interviewee's experiences and perspectives"* (Mohajan & Mohajan, 2022).

All four authors have been associated with Yim Tin Tsai in varying roles over 20–30 years so there is a strong element of longitudinal analysis that, we suggest, is relatively rare in many tourism research studies. Dr Sofield and Dr Li first visited Yim Tin Tsai in 1993 and 1994 as part of a consultancy team that developed a Tourism Strategy for the Sai Kung District (Ap et al., 1994). That report included *inter alia* a recommendation that the abandoned salt pans of Yim Tin Tsai be restored as part of the cultural heritage attractions of the district. At that time there were still 10 households in the village, and the primary school and church were still functioning. Dr Marafa has been taking students from the Chinese University of Hong Kong's Master tourism programme on field trips to Yim Tim Tsai since 2001. And Kwo Fung Shek, who has been attending church services on the island since he was a child, was foundation director of the community's NGO, "Salt & Light", which was established in 2013 to oversee pilgrims' tours, and he was a member of its Executive Committee until 2021. In those positions he played a key role in the management of the island over a 5-year period and provides an emic perspective to our analysis. In addition to his constant activist/observer presence on the island as its visitation burgeoned, the other authors made a total of 15 field trips to Yim Tin Tsai over a 4-year period in 2018–2019 and in 2021–2023, approaching the situation from anthropological and ethnological standpoints where participant observation was combined with open-ended, often opportunistic, loosely-constructed interviews of key stakeholders, former residents of the island, volunteers and visitors. In an adjunct research project (*"Community-based Narratives and Public Experiential Engagement for Cultural and Historical Heritage Conservation and Revitalisation of Yim Tin Tsai, Sai Kung"*) by the Department of Geography & Resource Management of the Chinese University of Hong Kong that is scheduled to continue through 2024, more than 50 former residents of Yim Tin Tsai have so far been interviewed to build a collection of narratives about their former life and experiences of growing up on the island, and we draw upon their stories to provide an emic voice that opens up aspects of the Hakka *communitas* (2022). This compendium of collective memory represents utilization of the 'narrative approach' (Chang & Lin, 2022; Olsen et al., 2016) to ascertain what community members themselves say about their unique culture, beliefs, values and ethnicity. As noted previously, Hakka identity has been an important element in the evolution of Yim Tin Tsai over generations in order to arrive at its contemporary socio-cultural ethos.

Both historical and contemporary materials were accessed to supplement information gleaned from personal, first-hand immersion in social and cultural events on the island that included attendance at church services and arts events. Yim Tin Tsai continues to have no permanent inhabitants, but it represents a multidimensional study in how an abandoned village has been transformed into a dynamic place of community-based pilgrimage and heritage tourism where Hakka culture and Catholicism have played defining roles. In essence, we apply our concept of heri-ligion in a bottom-up approach through an ethnographic study in which ethnic, religious and secular constituents entangle.

References

Ap, J., Li, F. M. S., & Sofield, T. H. B. (1994). *The Saikung Region of Hong Kong: A Tourism Strategy*. Saikung District Council.

Bauman, Z. (1996). From Pilgrim to Tourist—Or a Short History of Identity. In S. Hall & P. Gay (Eds.), *Questions of Cultural Identity* (pp. 18–36). Sage.

Bogumil, Z., & Lukaszewicz, M. (2018). Between History and Religion: The New Russian Martyrdom as an Invented Tradition. *East European Politics and Societies, 32*(4), 936–963. https://doi.org/10.1177/0888325417747969

Bourdieu, P. (1993). *The Field of Cultural Production: Essays on Art and Literature*. New York: Columbia University Press.

Chang, C.-C., & Lin, Y.-H. (2022). Constructing Hakka Ethnic Identity Through Narrative Genealogy Writing. *SAGE Open*, *12*(1). https://doi.org/10.1177/21582440221079913

Charmaz, Kathy (2006). *Constructing Grounded Theory. A Practical Guide Through Qualitative Analysis*. London: Sage Publications.

Charmaz, K. (2014). *Constructing Grounded Theory: A Practical Guide Through Qualitative Analysis* (2nd ed.). Sage.

Ching, J. (1993). *Chinese Religions*. Orbis Books.

de Jong, F., & Mapril, J. (Eds.). (2023). *The Future of Religious Heritage: Entangled Temporalities of the Sacred and the Secular*. Routledge.

Deschepper, J. (2018). Le «patrimoine soviétique» de l'URSS à la Russie contemporaine Généalogie d'un concept. *Vingtième siècle (Paris. 1984)*, *137*(1), 77–98. https://doi.org/10.3917/ving.137.0077

Di Giovine, M. A., & Garcia-Fuentes, J. M. (2016). Sites of Pilgrimage, Sites of Heritage: An Exploratory Introduction. *International Journal of Tourism Anthropology*, *5*(1–2), 1–23.

Di Giovine, M. A. (2012). Padre Pio for sale: Souvenirs, Relics, or Identity Markers? *International Journal of Tourism Anthropology*, *2*(2), 108–127.

Di Giovine, M. A. (2021). Religious and Spiritual World Heritage Sites. In M. A. Olsen & D. Timothy (Eds.), *The Routledge Handbook of Religious and Spiritual Tourism* (ch.15). Routledge.

Di Giovine, M. A., & Choe, J. (2019). Geographies of Religion and Spirituality: Pilgrimage Beyond the "Officially" Sacred. *Tourism Geographies*, *21*(3), 361–383. https://doi.org/10.1080/14616688.2019.1625072

Douglas, A. (1994). *Symbiotic Interactions*. Oxford University Press.

Douglas, A. E. (2010). *The Symbiotic Habit*. Princeton University Press.

Edgley, A., Stickley, T., Timmons, S., & Meal, A. (2016). Critical Realist Review: Exploring the Real, Beyond the Empirical. *Journal of Further and Higher Education*, *40*(3), 316–330. https://doi.org/10.1080/0309877X.2014.953458

Farias, M., Coleman, T. J., Bartlett, J. E., Oviedo, L., Soares, P., Santos, T., & Bas, M. del C. (2019). Atheists on the Santiago Way: Examining motivations to go on pilgrimage. *Sociology of Religion*, *80*(1), 28–44.https://doi.org/10.1093/socrel/sry019

Foy, G. (2023). *Chinese Religions and Philosophies: From Past to Present and Present to Past*. Asia Society. https://asiasociety.org/chinese-religions-and-philosophies

Graburn, N. (2001). Secular Ritual: A General Theory of Tourism. In V. Smith & M. Brent (Eds.), *Hosts and Guests Revisited: Tourism Issues of the 21st Century*. Cognizant Communications.

Graburn, N. H. H. (1977). Tourism: The Sacred Journey. In V. Smith (Ed.), *Hosts and Guests: The Anthropology of Tourism* (pp. 17–31). University of Philadelphia Press.

Harvey, D. C. (2015). Heritage and Scale: Settings, Boundaries and Relations. *International Journal of Heritage Studies: IJHS, 21*(6), 577–593. https://doi.org/10.1080/13527258.2014.955812

Hirsch, F. (2014). Empire of Nations: Ethnographic Knowledge and the Making of the Soviet Union. *Cornell University Press.* https://doi.org/10.7591/978 0801455940

Hodder, I. (2023). *Entangled: A New Archaeology of the Relationships Between Humans and Things.* Wiley.

Home Affairs Department, Government of Hong Kong Special Administrative Region. (2023). *Rural Representative Elections.* https://www.had.gov.hk/rre/eng/index.htm

Hui, Y.-F., Hsiao, H.-H. M., & Peycam, P. (2018). Introduction: Finding the Grain of Heritage Politics. In H. Hsiao, Y. Hui, & P. Peycam (Eds.), *Citizens, Civil Society and Heritage-Making in Asia* (Ch.1). ISEAS Publishing.

Lin, Z. (林作尧). (2022, November 17). 客家人"根"在河洛，文化源远流长 (Hakka Culture). https://mp.weixin.qq.com/s/VpZjTa58RYu5DhGJFf hmSQ

Luehrmann, S. (2015). *Religion in Secular Archives: Soviet Atheism and Historical Knowledge.* Oxford University Press.

Martin, B. D., & Schwab, E. (2012). Symbiosis: "Living Together" in Chaos. *Studies in the History of Biology, 4*(4), 7–25.

Meyer, B., & de Witte, M. (2013). Heritage and the Sacred: Introduction. *Material Religion, 9*(3), 274–280. https://doi.org/10.2752/175183413X13730 330868870

Mohajan, D., & Mohajan, H. (2022). *Constructivist Grounded Theory: A New Research Approach in Social Science* (MPRA Paper No. 114970), posted online 15 October 2022. https://mpra.ub.uni-muenchen.de/114970/

Murray, M. (2021). *The Camino de Santiago: Curating the Pilgrimage as Heritage and Tourism.* Berghahn Books.

Niedźwiedź, A., & Saraiva, C. (2015, June 22). *The Heritagization of Religious and Spiritual Practices: The Effects of Grassroots and Top-Down Policies.* SIEF Ethnology of Religion Working Group conference. Zagreb: International Society for Ethnology and Folklore. https://www.nomadit.co.uk/sief/sief2015/panels.php5?PanelID=3394

Olson, B. D., Cooper, D. G., Viola, J. J., & Clark, B. (2016, 1 January). Community Narratives. In L. A. Jason & D. S. Glenwick (Eds.), *Handbook of Methodological Approaches to Community-Based Research: Qualitative, Quantitative, and Mixed Methods* (Online ed.). Oxford Academic. https://doi.org/10.1093/med:psych/9780190243654.001.0001

Oxford English Dictionary (2012). London: W.H. Smith.

Peycam, P., Shu-Li, W., Yew-Foong, H., Hsin-Huang Michael, H. (Eds.). (2020). *Heritage as Aid and Diplomacy in Asia*. ISEAS—Yusof Ishak Institute; International Institute for Asian Studies (IIAS); and Institute of Sociology, Academia Sinica.

Pimenova, K. (2022). Museums and Religious Heritage: Post-colonial and Post-socialist Perspectives. *Civilisations Revue Internationale d'anthropologie et de Sciences Humaines, 71*, 13–28.

Sagarin, R. (2013, June 25). *To Overcome Your Company's Limits, Look to Symbiosis*. Harvard Business Review. https://hbr.org/2013/06/to-ove rcome-your-companys-limits-look-to

Santos, X. M. (2002). Pilgrimage and Tourism at Santiago De Compostela. *Tourism Recreation Research, 27*, 41–50.

Sofield, T. H. B. (2013). Angkor: Tourism Management Caught in the Crossfire. In M. P. Oton, P. J. M. Mantinan, & V. P. Carril (Eds.), *Touristic Management of World Heritage Monuments and Cities* (pp. 117–152). Universidade de Santiago de Compostela Publications.

Thouki, A. (2022). Heritagization of Religious Sites: In Search of Visitor Agency and the Dialectics Underlying Heritage Planning Assemblages. *International Journal of Heritage Studies: IJHS, 28*(9), 1036–1065. https://doi.org/10.1080/13527258.2022.2122535

Timothy, D. J. (2011). *Cultural Heritage and Tourism: An Introduction*. Chanell View Publications.

Timothy, D. J. (2021). *Cultural Heritage and Tourism: An Introduction* (2nd ed.). Channel View Publications.

Torraco, R. J. (2016). Writing Integrative Literature Reviews: Using the Past and Present to Explore the Future. *Human Resource Development Review, 15*(4), 404–428. https://doi.org/10.1177/1534484316671606

Turner, V. (1974a). *Pilgrimage and Communitas. Studia Missionalia, 23*, 305–307.

Turner, V. (1974b). *Dramas, Fields, and Metaphors: Symbolic Action in Human Society*. Cornell University Press.

Wang, S. L., Rowlands, M., & Zhu, Y. (Eds.). (2021). *Heritage and Religion in East Asia*. Routledge.

CHAPTER 3

Yim Tim Tsai—The Place and Its History

Abstract This chapter outlines a brief history of Yim Tin Tsai over 370 years from the time of its original settlement by the migrating Hakka Chan clan from mainland China to the current day. It charts its transformation from Daoist-oriented Hakka folklore as its religious foundations to Catholicism which the community adopted in the mid-1800s. It sets out the brief but crucial connection of Father Joseph Freinademetz to Yim Tin Tsai and his subsequent missionary work in China which led to his canonization as a saint by the Pope in 2003. This in turn led to the revitalization of Tim Tin Tsai from a totally abandoned village to a thriving pilgrimage site as the Village Committee and the Catholic Diocese seized 'ownership' of the new saint and used the Vatican's decision about him to restore the 1889 St Joseph's chapel. Through a Joint Committee they established a not-for-profit agency called "Salt and Light" to oversee the development of the island, which is now evolving into a broader, 'secular' tourism destination with a distinct identity.

Keywords Chan clan · Solar evaporation · Salt-making · Hakka folklore · Catholicism · Father Joseph Freinademetz · Canonization · Co-option · Restoration of built heritage

T. Sofield et al., *Heritage-Making in Hong Kong Through Culture and Religion*, https://doi.org/10.1007/978-981-97-4339-1_3

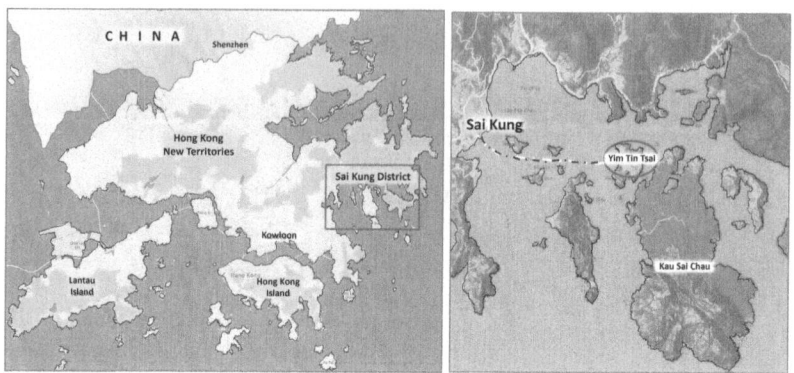

Diagram 3.1 Maps of Hong Kong showing the location of Yim Tin Tsai (*Source* Department of Geography and Resource Management, Chinese University of Hong Kong, as modified by Sofield)

Yim Tim Tsai is a small island of less than 1 sq.km that lies in the northeast of the New Territories of Hong Kong, two kilometres off the coast from Saikung (see Diagram 3.1). According to Ticozzi (2008, cited in Atha, 2014), genealogical records indicate that the island was settled in 1751 by a Hakka clan with the single-family name Chan from the village of Yim Tin in Yantian district of south-eastern Guangdong, mainland China, its founder identified as Chan Mang Tak. This date correlates well with the youngest current members of the clan being 12th generation (Atha, 2014, p. 11). The settlement was representative of thousands of Hakka villages whose community society is structured on a patrilineal kin group with one common surname. The Yantian Hakka community probably chose to settle on Yim Tin Tsai because it had the key topographical feature needed for their traditional cottage industry, salt making—a flat area that was flooded by the sea at high tide and which drained at low tide, where they could construct salt pans utilizing the sun to evaporate the water (Lin 1940; Hase & Se Yan, 2009). The mangrove forests edging the seashore were also an essential feature because they could protect the fields from typhoons and strong waves and act as a filter through which the saltwater flowed. At high tide, a gated sluice channel brought saltwater into the fields through the mangrove forest. The island's name is derived from this activity since Yim Tin Tsai means "Little Salt Field" (Lin, 1940) (see Plate 3.1).

Plate 3.1 Aerial photograph of Yim Tin Tsai (*Source* Hong Kong Tourism Commission with annotations by the authors)

Originally the village farmed 6 acres of salt pans and together with a little agriculture on the adjoining island of Kau Sai Chau and fishing in the surrounding waters was relatively prosperous. The opening of the Kowloon Canton Railway (KCR) in 1910 marked the death knell for Yim Tin Tsai's cottage salt industry, however, as cheaper and much larger volumes of salt were imported from China. The salt pans were abandoned within a year and the villagers transferred their energies solely to farming and fishing. But by the 1970s the island economy was in dire straits and the majority of families moved first to mainland Hong Kong and then many of them overseas. By 1995 only two families were still living on the island, and both the church and the Ching Po primary school were closed and in semi-ruins. By the turn of the century the island was entirely abandoned, its houses derelict, in Ashworth's term, a "*relict built environment*" (2011, p. 1) (see Plate 3.2).

Plate 3.2   A 'relict built environment'—The 'Ghost Village' of Yim Tin Tsai

3.1   Conversion from Hakka Folklore to Catholicism

Following the annexation of Hong Kong by the British Government in 1842, the Vatican assigned Catholic missionaries to the new Crown colony. In 1866, two Italian priests, Father Simeon Volonteri and Father Gaetano Origo, were given tutelage over the District of Saikung, which includes the island of Yim Tin Tsai. By 1876 they had converted the entire village to Catholicism: it has been classified officially as a Catholic village ever since (Ticozzi, 2008). Its first chapel (now in complete ruin with only parts of three crumbling walls remaining) was converted from a village house in 1879 (Atha, 2014) for a young Austrian missionary, Father Joseph Freinademetz, who lived on the island for two-and-a-half years to learn Hakka Chinese before being transferred by his order, the Society of Divine Word Missionaries, to Shandong Province in mainland China in 1881. The little chapel was replaced with a much larger building (St Joseph's Chapel) constructed in the Italian Romanesque Revival style in 1889–1890. Father Freinademetz was a co-founder of the Society's work in mainland China and he died in southern Shandong on 28 January 1908 (Catholic Heritage Organization, 2016).

In the 1970s, after most of the villagers relocated to mainland Hong Kong and overseas to Britain, Canada, Australia and elsewhere, the church slowly deteriorated in the absence of constant use and maintenance. Some of the local villagers organized an annual mass on St Joseph's Feast Day in May every year as one way of trying to maintain community contact with their birthplace, but otherwise the chapel was locked up. At the turn of the century, the deserted village was utilized by a recreation company for its customers to conduct 'paintball wars' (where two teams armed with modified guns shoot balls of paint at each other) and the church and village houses were vandalized. The Catholic Diocese of Hong Kong had the de-consecration of the church under consideration because the cost of repairs and maintenance in the absence of an active congregation was considered a poor use of funds.

At this point, the Vatican in Rome became a crucial catalytic agency. Father Freinademetz, the long-dead nineteenth-century missionary, was the unwitting stimulus for the intervention by the Vatican that led to Yim Tin Tsai's embrace of an exercise in religious heritagization and co-creation.

3.2   The Canonization of Father Joseph Freinademetz and the Resurrection of Yim Tin Tsai through Pilgrimage

Some 80 years earlier, in 1934 the then Superior General of the Society of the Divine Word Missionaries, Father Joseph Grendel, had proposed to the Vatican's Committee of Canonization that Father Freinademetz, who had been venerated on his death in 1908, be accorded sainthood (Hollweck & Ueblackner, 2008). Veneration is the first of three steps culminating in canonization as a saint, with beatification as the second step. Father Freinademetz had become very well known throughout the Catholic world because of his work in China (e.g. Henninghaus' *magnus opus* of 620 pages, 1920; Bornemann, 1926; Aulitzky, 1932; Fischer, 1936; Baur, 1939, all cited in Steffan, 2012). Three major reasons lay behind the submission:

(i) Recognized as a holy, devout Christian, a number of miracles of healing, and saving lives during the Boxer Rebellion (Nov 1899–Sept 1901), were attributed to Father Freinademetz;

(ii) His profound understanding of the Chinese language and Chinese customs allowed him to communicate Christian theology in ways that bridged differences rather than dismissing traditions arising from Daoism, Confucian thought, ancestor worship, *feng shui* practices, etc., as 'pagan'. He pioneered a quasi-syncretic approach that was adopted globally by his Order in its missionary work; and

(iii) He was considered to be mainly responsible for converting more than half a million Shandong Province people to Catholicism (Steffan, 2012).

The investigation by Father Grendel's team lasted almost four years but the outbreak of the Second World War sidelined the process. However, the Divine Word Missionaries Society held their former colleague in such high esteem that in 1950 they bought the Freinademetz family home in the hamlet of Oies, Alta Badia. (Originally in the area of South Tyrol in Austria, it is now in northern Italy following adjustments to the border between Austria and Italy after the First World War.) Initially a monastic retreat for their own missionaries it quickly became a pilgrimage site and in 1962 the Society constructed a church in the village for pilgrims (Hollweck & Ueblackner, 2008). In 2019 about 20,000 pilgrims visited the village (Suedtirolerland.it, accessed 14 Feb 2021). In 1975, the second of the three stages in the process of canonization occurred when Father Freinademetz was beatified on World Mission Day celebrated by the Catholic Church, i.e. he was declared "Blessed" and merited public worship. At the turn of the twentieth century the case for Father Freinademetz to be elevated to sainthood gained renewed vigour and in May 2003 by papal decree he was canonized (John Paul II, 2003). His elevation to sainthood was based on Vatican-endorsed esteem for his work in China. As such the Vatican had no particular concern for or interest in Yim Tin Tsai. However, for the former village community of Yim Tin Tsai and for the wider Catholic community of Hong Kong in general, the Vatican's move was highly symbolic. Importantly, its significance for both the island community and Hong Kong was recognized by the Vicar General of the Catholic Diocese of Hong Kong, Father Dominic Chan Chi-ming, an eighth generation Yim Tin Tsai islander. The ruins of Father Freinademetz's little chapel were directly behind the Vicar General's ancestral home, and the priest's cottage was located just four doors away. The canonization gave the village community and the broader Hong Kong Catholic community a claim to its first saint, automatically accompanied with a degree of global exposure. They immediately exercised 'ownership' of him in a partnership with the Village Committee by which they established the Joint Organizing Committee of Yim Tin Tsai, and then launched a non-governmental fund-raising drive to restore the then run-down church. The trajectories of both stakeholders in reaching agreement were different but the outcome coincided: a greater visibility, status and prestige for the Catholic church, and a path to restoration of Yim Tin Tsai for the village community—a clear example of mutualism, i.e. a symbiotic relationship from which both parties benefit.

HK$1 million (USD$150,000) was raised. As part of the restoration of the church, successfully completed in 2004 with significant volunteer input, the Catholic community commissioned a statue and a stained-glass window of Father Freinademetz, now installed in the church. In 2005, UNESCO honoured the restoration endeavour with an 'Award of Merit' under its Asia–Pacific Awards for Culture Heritage Conservation (see Plate 3.3). In 2006, on the first Sunday of May, Vicar General Chan Chi-Ming held a mass in the newly-restored church to celebrate St Joseph's Feast Day. Every year since on the same anniversary (until the advent of COVID-19 restrictions), hundreds of pilgrims attend the celebratory mass. In 2022, with some pandemic restrictions relaxed, the celebration of St Joseph's Feast Day was held on 23 October, with mass again celebrated by Vicar General Chan. In the past this event attracted many of the former village residents both locally and a number from overseas who also returned for an annual reunion (Chan, Colin Chung-yin, President, Yim Tin Tsai Rural Village Committee, personal communications; and Kwo Fung Shek, personal communications, 2018). However, for the 2022 church service, overseas attendees were quite small in number since COVID-19 entry restrictions to Hong Kong remained onerous and airfares very costly, as fieldwork observations by Sofield & Shek in October 2022 revealed. The May 2023 celebration of St Joseph's Feast Day witnessed more than 200 local celebrants but about only 15 from overseas.

Following the co-option of Father Freinademetz as 'their saint' by the Yim Tin Tsai community the church quickly became a site for Catholic pilgrimage throughout all months of the year. Visitation began to expand substantially and as noted in 2013 the HK Catholic Church and the Yim Tin Tsai Village entered into a formal partnership through the Joint Committee to create their NGO called "Salt & Light". It was charged with establishing religious-based interpretation and tours, managing all visitation, including arranging village-owned *taiko* (small boats converted to ferry passengers), and fund-raising, again a further example of mutualism. Before the onset of the COVID-19 global pandemic in 2019 an estimated 300,000 visitors had toured the island between 2005 and 2018, of whom more than 250,000 were pilgrims ("Salt & Light" administrator, personal communication, 2019). Detailed statistics on visitor numbers maintained by "Salt & Light" from its inception in 2013 to 2018 recorded a total of 181,655 of whom 68,209 were listed as individual pilgrims (although many were families, small groups of friends,

Plate 3.3 St Joseph's Chapel, restored

etc.), some **99,446** came as registered pilgrim groups for whom formal tours were organized, (both of these categories arriving on village *taiko*), and an estimated 14,000 were "itinerant tourists" who arrived unheralded on "other boats". In 2016, which was designated by the Vatican as the International Holy Year of Mercy (Jubilee Year), some 37,108 pilgrims were welcomed by "Salt & Light", for whom the church was the focal point of their trip to Yim Tin Tsai.

The impact of the lockdowns caused by the COVID-19 pandemic saw visitation to Yim Tin Tsai plummet, but once restrictions were lifted in 2021 and "Salt & Light" moved to a six-day week instead of weekends only, visitation soared. In 2022 total visitation was estimated at 80,000 and for 2023 the total number was expected to reach 100,000 (Shek, personal communication, December 2023).

3.3   The Restoration of the Salt Pans

The stimulus for the restoration of the salt pans eventuated via a serendipitous trajectory. In 2011 the Joint Catholic/Village Committee commissioned a Hong Kong film maker, Wong Tin-shing, to make a documentary of the conservation of the church, incorporating also the then recent refurbishment of the primary school as a folk museum (Wong, 2013). Before long, fascinated by the 'other' key element of Yim Tin Tsai's heritage, i.e. traditional salt-making, that was being largely over-looked because of the emphasis on Catholicism, Wong's enthusiasm transpired into advocating efforts to restore the salt pans. He was able to gain the backing of the Joint Organizing Committee with strong support from the architect who had supervised the conservation of the church, Anna Kwong. No remaining descendants had any knowledge of the traditional salt-making industry utilizing evaporation of shallow pools of salt water by the sun, and so a team of volunteers was dispatched to mainland China and to Taiwan where examples still existed, to gain first-hand knowledge of the engineering and processes involved. Another fund-raising effort was launched, some HKD$6 million was raised, the overgrown mangroves were cleared from the mudflats with a combination of volunteers and commercial input, they were drained, bunds erected and by 2014 three acres had been restored. The aim was to re-establish the salt pans not as a commercial operation but rather with a three-fold aim: to restore the traditional way of salt making to demonstrate the process to the public, to portray its associated culture as a Hakka cottage industry, and to promote sustainable development using passive energy (i.e. solar evaporation). The intention of the project was to highlight six distinct but interrelated features of Yim Tin Tsai: religion, culture, heritage, ecology, tourism and education/interpretation, with weekly guided tours around the site. The combination of sacred heritage (the church and associated ceremonies) and industrial heritage initially proceeded smoothly, with a tour of the salt fields seen as an adjunct to the pilgrimage-focused tours. This development may be seen as a form of commensalism which added to the richness of a religiously-oriented visit to Yim Tin Tsai, captured in the title of the NGO, "Salt & Light" with its strong Biblical connotations, and the secular nature of the salt-pans restoration not detracting from the original intent of pilgrimage tours.

Adhering as far possible to the original materials with which the salt pans had been constructed, the restoration project included five main components: a sluice gate and ingress channel, a large holding pond to catch and contain seawater at hightide, four evaporation pools through which the seawater flows in sequence, becoming more and more salty with each subsequent pool, a brine pond and a crystallization pool from which the pure salt is harvested (see Plate 3.4). UNESCO again recognized the community's endeavours when in October 2015 the effort was granted UNESCO's 'Asia–Pacific Award for Industrial Heritage Conservation' (Chapman, 2020; UNESCO, 2020; UNESCO, 2020). In 2022 the Hong Kong Health authorities assessed the quality of the salt as high and fit for human consumption, so that small packets are now sold as souvenir items.

UNESCO's second intervention also opened the door to a perception of Yim Tin Tsai as a more generalized heritage site that tended to downplay the sacralization that dominated its original restoration efforts.

Plate 3.4   The restored salt pans of Yim Tin Tsai, with the village and church overlooking the site

With the advent of the Hong Kong Tourism Commission as a major new stakeholder, the objectives of the Catholic Church diverged from those of the Commission, with the Village Committee displaying signs of division as some supported the move towards increased tourism and others remained wedded to the emphasis on pilgrimage visitation focused on sacralization. This divergence of opinion and objectives has inevitably led to a degree of competition, so that dissonance and entanglement (in the classic sociological context) entered the situation. The salt pans have become a popular attraction for most visitors in addition to the religious heritage sites; and as an education hub for schoolchildren who are able to participate in different parts of the salt-making process it is further evidence of co-creation activities on Yim Tin Tsai.

REFERENCES

Ashworth, G. J. (2011). Preservation, Conservation and Heritage: Approaches to the Past in the Present Through the Built Environment. *Asian Anthropology, 10*(1), 1–18. https://doi.org/10.1080/1683478X.2011.10552601

Atha, M. (2014). *Archaeological Survey-Cum-Excavation, Yim Tin Tsai, Sai Kung (Oct.–Nov. 2013)*. Hong Kong Archaeological Society.

Aulitzky, J. M. (1932). *Kurzes Lebensbild des P. J. Freinademetz*. Mödling

Baur, J. (1939) *Der Diener Gottes P. Joseph Freinademetz SVD. 1852–1908. Das Leben eines heiligmäßigen Chinamissionärs dem Volke erzählt, Missionari Verbiti*. Varone.

Bornemann, F. (1926). *As Wine Poured Out. Blessed Joseph Freinademetz. Missionary in China 1879–1908* (Fr. J. Vogelgesang, Trans., 1984). Divine Word Missionaries.

Catholic Heritage Organization. (2016). *St Joseph's Chapel, Yim Tin Tsai*. https://www.catholicheritage.org.hk/en/catholic_building/yim_tin_tsai/index.html

Chapman, W. (Ed.) (2020). *Asia Conserved, vol. IV: Lessons Learned from the UNESCO Asia-Pacific Heritage Awards for Culture Heritage Conservation, 2015–2019*. Southeast University Press. https://unesdoc.unesco.org/ark:/48223/pf0000374413

Fischer, H. (1936). *Joseph Freinademetz. Steyler Missionary in China. Ein Lebensbild (A Biography)*. Missionsdruckerei (Missions Printing Office).

Hase, P. H., & Yan, S. (2009). The Price and Consumption of Salt in China in 1901. *Journal of the Royal Asiatic Society Hong Kong Branch, 49*, 127–218.

Henninghaus, A. (1920). *P. Jos. Freinademetz S.V.D. Sein Leben und Wirken. Zugleich Beiträge zur Geschichte der Mission in Süd-Schantung*. Verlag der Katholischen Mission.

Hollweck, S., & Ueblackner, S. (2008). *Pioneer of the Divine Word Missionaries in China—Joseph Freinademetz Serving the People of China.* https://www.svd curia.org/public/histtrad/founders/jf/jfen.htm

Lin, S. Y. (1940, January). Salt Manufacture in Hong Kong. *The Hong Kong Naturalist,* Vol X(1), pp 14–20.

Pope John Paul II. (2003, October 5). Canonization of Three Blesseds. *Libreria Editrice Vaticana.* https://www.vatican.va/content/john-paul-ii/ en/homilies/2003/documents/hf_jp-ii_hom_20031005_canonizations.html

Steffen, P. (2012). Witness and Holiness, the Heart of the Life of Saint Joseph Freinademetz of Shandong. *Studia Missionalia, 61,* 257–392.

Ticozzi, S. (2008). The Catholic Church and Nineteenth Century Village Life in Hong Kong. *Journal of the Hong Kong Branch of the Royal Asiatic Society, 48,* 111–149.

UNESCO. (2020). *Asia-Pacific Awards for Cultural Heritage Conservation, 2015–2019* (Vol. IV). UNESCO. https://unesdoc.unesco.org/ark:/48223/ pf0000374413.pdf. Accessed 22 November 2020.

Wong, T. S. (Dir.). (2013). *Then and Now—Yim Tin Tsai: A Hakka Catholic Village Reborn* [Documentary film]. New Wave Studios.

CHAPTER 4

'Heri-ligion': The Heritagization of Religion and the Sacralization of Heritage

Abstract In this chapter, we survey the rich tapestry of numerous processes, events and interrelationships that illustrate 'heri-ligion' in practice in Yim Tin Tsai as defined by the eleven factors listed in Chapter 2 that constitute our concept. The transformation of Yim Tin Tsai, from a village originally predicated geographically and sociologically on the application of feng shui to a community guided by Catholicism, is charted. Sacralization and secularization are evident as major components of heritage-making and are manifested in many configurations.

Keywords Heri-ligion · Sacralization · Secularization · Sacred/secular gaze · Heritagization · Daoism · Geomancy · *Feng shui* · *Chi* · Chinese concept of harmony · "Catholic Nature Trail of Reconciliation"

In this chapter, we survey the rich tapestry of numerous processes, events and interrelationships that illustrate 'heri-ligion' in practice in Yim Tin Tsai as defined by the eleven factors listed in Chapter 2 that constitute our concept. The processes associated with heri-ligion are exemplified in the initial resurrection of Yim Tin Tsai from an abandoned village to a dynamic pilgrimage site which now also attracts ever-increasing numbers of secular tourists. In the process, the multifaceted Hakka world view that combined elements of Buddhism, Doaism, ancestral worship, Hakka

© The Author(s), under exclusive license to Springer Nature Singapore Pte Ltd. 2024
T. Sofield et al., *Heritage-Making in Hong Kong Through Culture and Religion*, https://doi.org/10.1007/978-981-97-4339-1_4

folklore and Confucian thought, was replaced with Catholicism so that the heritage of Yim Tin Tsai has the one overlaid on the other, with the ethnic identity of the Hakka community the 'glue' that has held fast to some aspects of traditional culture while others have been modified. And most recently, commercialization and commoditization of heritage have been superimposed over the sacred, thus leading to a 'sacred/secular gaze' with duality of interpretation.

St Joseph's Chapel represents the dominant feature of heri-ligion on the island, and its restoration and conservation constitute: "*a catalyst project for enhancement of the cultural landscape of Yim Tin Tsai*" (Chan et al., 2012). It encapsulates the four main attributes of 'sacrosanct' built heritage forms: externals (i.e. architecture), internal (i.e. images), eternal (i.e. universal message), and manifestive (i.e. adherents' beliefs), as identified by Singh (2008). The Chapel dominates the village, its commanding hilltop location and orientation determined by its rejection of the basic principles of Daoist landscape design to be a demonstration of the power and superiority of Christianity over what the Catholic Church regarded as a pagan religion.

In this context, the village was originally laid out according to traditional *feng shui* principles, where the buildings are aligned on a north-south orientation to capture *chi*, the all-powerful force that emanates from the north Polar Star. *Feng shui* is integral to Daoism and is a synthesis of Chinese traditional cosmology, geomancy, astrology, geography and numerology. This synthesis was stylized through a chart of the cosmos, which placed man, state, nature, and heaven in harmony. The philosophy, precepts and principles of *feng shui* were first comprehensively detailed by the Duke of Zhou who compiled a famous Chinese classic on its application called the "*Zhou Li*" ("*Rites of Li*"), circa 900 B.C. (Sofield et al., 2017). Since the endeavours of humans are subjected to the twin influences of Heaven and Earth, *feng shui* was designed to provide a mechanism by which the *yin* and the *yang* of *chi* could identify places where the forces of Heaven and Earth would be in harmony. Underground, the forces of *chi* flow through dragon's veins; above, they manifest chiefly as wind (*feng*) and water (*shui*)—the term *feng shui* means "wind" and "water", and *sha-qi*, inauspicious (evil) energy forces emanating from natural landforms. Thus, crucial to *feng shui* are features in the landscape. For example, mountains (*shan*) are *yin*, meaning passive, with ascribed characteristics of the dragon, tiger, turtle or phoenix. They could be balanced by water which is *yang*, meaning active, able to attract and hold wealth. And their juxtaposition would determine the flow of forces or energy between them and whether a site was to be avoided or developed for a particular purpose—traditionally a town, a temple, a palace, a residence, a grave and now extended to include the premises of a business, a community facility, etc. (Sofield et al., 2017).

Through the application of geomancy, auspicious sites could be identified where a building or town could benefit from the energized spot called the *xue*, literally translated as 'the dragon's lair', where the *chi* collects/focuses and is the best possible site for human ingress into a landscape. For more than 2000 years, for example, every imperial palace in China was located in/on the *xue* of a site identified by geomancers as the most favourable place in the empire to bestow *chi* upon the emperor, whose over-riding duty was to ensure harmony between heaven, earth and his subjects. The application of *feng shui* reflects the notion that: "*Human alterations of the landscape do not simply occupy empty space. Rather, sites are viewed as manifesting certain properties which influence, even control, the fortunes of those who intrude upon the site*" (Knapp, 1986, pp. 108–109).

The last section of the *Zhou Li*, the "*Kao Gong Ji*" (*Book of Diverse Crafts*), encapsulates those elements of *feng shui* that ideally should be applied in locating man-made constructions of all kinds right down the scale from cities to individual buildings in a landscape to ensure harmony between the heavens, earth and humans. Thus, the site should be:

- oriented along a north-south axis in order to capture the *chi* that emanates from the North Polar star;
- built on a south-facing slope to benefit from the warmth of the sun in winter shining on the face of the buildings and on the back of the buildings in summer [passive energy];
- flanked by the arms of encircling hills, viz—
  - the dragon range to the north to protect the site from the chill Arctic winds,
  - a slope immediately behind the site called the 'black tortoise ridge' to provide protection from *sha-qi*, covered in a *feng shui* forest that is designated as protected and semi-sacred. Famously, an artificial hill (called *Jingshan*) was created at the northern extremity of the Forbidden City in Beijing because it lacked a natural black tortoise ridge;

- the white horse hill on the west to protect the site from westerly *sha-qi*;
- the green tiger hill to the east for protection from easterly *sha-qi*;
- the phoenix range to the south to lessen the dangers from *sha-qi*; and
- a lake or meandering river in front of the village to provide a steady water supply and ensure that wealth accumulates, in contrast to a swiftly flowing river which is characterized as carrying wealth away in floods (Li, 2008).

For almost three thousand years, settlements in China were planned in the context of this cosmic framework to maintain harmony and balance, developing the prototype of urban planning for literally millions of sites and buildings that was followed down the centuries and is still utilized to some extent in modern China. In Hong Kong, *feng shui* is commonly applied (Bai, 2003). The practices of *feng shui*, when isolated from Daoist beliefs, myths and legends, are in fact consistent with principles of conservation and good land management (Williams & Webb, 1994). For example, the Black Tortoise Ridge was covered in a *feng shui* forest (both natural and 'improved' with fruit and medicinal trees and shrubs) that was designated as semi-sacred and fully protected by village elders, thus reducing risks of erosion and landslides; the White Horse Hill sheltered the site from the westering sun; the Green Tiger Hill lessened the impact of typhoons sweeping in eastwards from the South China Sea; and similarly the Phoenix Range reduced the impact of the annual monsoon rains as they moved across the country from the south. These planning principles pre-date by two-and-a-half millennia some of the elements of the contemporary notion of ecologically sustainable development (ESD), especially in terms of site placement for passive energy (warming and cooling), protection from the elements, minimization of the effects of human construction in terms of erosion and ecological degradation, and environmental risk reduction (Sofield et al., 2017).

Many Hakka communities applied geomancy to locate their villages and guide the design and placement of buildings within the landscape, and there are more than 30 such examples located around the New Territories of Hong Kong where they settled, e.g. Lai Chi Wo and So Lo Pun, as well as Yim Tin Tsai. In a comprehensive survey, Chu (1998) listed 112 *feng shui* settlements and woodlands around the New Territories.

Plate 4.1 (a) Spatial layout of Yim Tin Tsai according to the ancient principles of "*feng shui*"; (b) St Joseph's Chapel constructed atop the Black Tortoise Ridge in 1889 (*Source* Hong Kong Tourism Commission)

When the Yim Tin Tsai community converted to Catholicism in the nineteenth century, a plot of land on the top of the Black Tortoise Ridge directly behind and above the village was donated by the elders for the building of a church. From this dominant location it asserts its spiritual superiority, an example of theologically-oriented cultural and colonial power. Placement on such a hilltop site is diametrically opposed to the traditional planning principles of *feng shui* where constructions are invariably located *within* the landscape. In effect this 'Daoist de-sacralization' of the village began a process of secularizing its surrounds and the current eighth generation of islanders no longer refer to any of its physical features as Daoist, and the labels that are generally applied to those features are absent in their discourse. Interestingly, however, some tourists familiar with the tenets of Daoism will 'read' the landscape in *feng shui* terms without necessarily acknowledging belief in its traditional Chinese religious foundations. Yim Tin Tsai's community and its mix of visitors thus exercise the sacred-secular gaze where their differing value orientations provide divergent interpretations; but there is no overt contention about the underlying contradictions, and commensalism is the result (see Plate 4.1).

There is one probably unintentional manifestation of Daoism in the actual orientation of the church. While the church building runs east to west with the altar located at the western (front) end and the main portal at the eastern end which is used for all main religious ceremonies, the most commonly-used entrance for weekly services and for visitors as both pilgrims and tourists instead of being at the back of the church as is traditional in many Christian churches (Cross & Livingstone, 2005), is located in the centre of the south side of the building. This entrance is approached by a wide set of steps that rise up the hill from the village along the north-south axis. The layout is a relatively subtle example of syncretism: in Daoist terms, the heavenly force, *chi*, is enabled to enter and empower the church through the strategic placement of the door although, as Catholics, the locals reject this notion.

To appreciate the full significance of the symbolism involved in siting the church atop the Black Tortoise ridge and its former *feng shui* forest, the Hakka Lutheran Protestant village of Shung Him Tong, near Fanling in the New Territories of Hong Kong, provides a contrasting outcome. When it tried to site its church on the top of the hill behind the village, violence ensued. Established as a farming community in 1903, its Village Committee decided to build a new, larger church on the hilltop in 1964 because, according to a village elder, *"it would be visible from far away and closer to non-Christian villages"*, thus calling them to Christianity (Constable, 1994, p. 124). However, the plan to construct the church above the village was perceived by the adjacent non-Christian, non-Hakka, Cantonese Daoist village of Lung Yeuk Tau as:

> a literal and figurative attempt by Christians to put themselves 'above' their neighbors and to prosper or benefit at the expense of the non-Christian village. ... The Lung Yeuk Tau villagers were against the church on the hill ... because they thought the hill was the source of their "feng-shui" and that the church would block it. ... A large group of people from the village organized a roadblock and threw stones to prevent the construction team from passing. (After several weeks of violent protests), the Christians abandoned the hill site and built the new church next to the old one on lower land. (Constable, 1994, p. 124)

The villagers of Shung Him Tong and Lung Yeuk Tau village shared a similar view about the symbolism of siting a religious building above the villages although their interpretations of that significance were diametrically opposed. Both communities considered that the effort was intended to demonstrate the ascendency of Christianity over Daoism—positively perceived by the Christian Protestants of Shung Him Tong but negatively perceived by the Daoists of Lung Yeuk Tau.

If any Daoists had remained among the Yim Tin Tsai community, then the twenty-first-century-Catholic-dominated interpretation of the heritage of the village with its absence of acknowledgement of the Daoist and *feng shui* roots of the community might have produced contestation (conceptualized as *"dissonance"* by Graham et al., 2000, p. 3). However, the monolithic conversion of all community members to Catholicism 160 years earlier had expunged any such notion and thus, to the extent that Graham et al. (2000, p. 4) claim that dissonance is always present as *"an intrinsic quality of heritage"*, the Catholic-oriented restoration of heritage by the present generation from Yim Tin Tsai tends to refute that notion. We suggest that the generalization by Graham et al. (2000) is too broad and that scale and socio-cultural uniformity are fundamental factors in determining whether or not dissonance will be manifest in the acceptance of heritage identification and interpretation. In a small, single-name clan village where all members of the community are related and adhere to the one religious belief, the scope for dissonance is limited, in contrast to a region or a multicultural nation where heritage symbols may be used in an attempt to unify sentiment for political and cultural purposes that will often invoke challenge and reaching consensus will often be contested.

Closely linked to the church and its ceremonial functions is the Catholic cemetery located on the back of a hill about 500 m from the village. It replaced the original Daoist grave sites on neighbouring Kau Sai Chau island. Given the fundamental role that acknowledgement of ancestors plays in Chinese and Hakka identity and attachment to place, this aspect of our study is explored in Section 8(iii) below on 'place attachment to home'.

In addition to the chapel, a 1.5 km long Nature Trail constructed in 2013 on Yim Tin Tsai presents a clear example of heri-ligion in action. It has taken eight 'secular' sites/locations and transformed them through sacralization into biblically relevant sites, with the whole trail designated as '*The Catholic Nature Trail of Reconciliation*' (Chan et al., 2012). It features as the central activity of a 2-hour biblically-oriented pilgrimage tour. Its theme is: *"To reconcile the frenzied lives"* of the busy, materialistic people of crowded Hong Kong who have become alienated from and destructive of Nature by *listening to the sounds of the peaceful environment of Yim Tin Tsai* and thus find a way *"to re-establish a harmonious relationship with people, Nature and God"* (Chan et al., 2012, p. 2).

The Nature Trail biblical tour starts with a prayer in the church, followed by visits to the 8 stations along the trail (each with relevant biblical passages). For example, the first stop is at a very old broad-leafed laurel tree on the hill near the church which symbolizes creation, with a

Plate 4.2   The Nature Trail of Reconciliation—Christianity and ecology combined; (a) Natural vegetation of Yim Tin Tsai; (b) Statue of Father Freinademetz along the Trail, located on the site of his original cottage (c) Station 5 on the Trail: grove of bamboos representing 'The Path and the Way to follow Christ'

verse from Genesis referencing God as the Creator of heaven and earth. This is followed by a short sermon (as are all the verses for each station) drawing out the teaching behind the chapter and verse. The village well, which had fallen in on itself many years ago, is now restored as the fourth station and sacralized with signage and a reading from the Gospel of John, ch.7:v37–39: "*Jesus stood and cried out: "Let anyone who is thirsty come to me! Let anyone who believes in me come and drink"*. In semiotic terms it becomes a signifier of Jesus and Christianity (Hall, 1997). The fifth station leads the pilgrim through a grove of bamboos, with a biblical quotation from the Gospel of John (ch.14:5–7) that refers to seeking the path ('the Way') to truth and life. The sixth station is located at the Catholic cemetery and references John ch.11 verses 25–26: "*Jesus said: I am the resurrection. Whoever believes in me even though he dies will live"*— and this verse could serve as a metaphor for the way in which the island has been brought back from an abandoned emptiness to life through its proselytization of Catholicism, sacralization of heritage, and pilgrimage. All eight stations along the Trail are similarly transformed from profane elements of Nature into sacralized Christian sites that are referenced biblically. The tour ends back at the chapel with hymns and another prayer. For secular tourists, the Trail immerses them in the sub-tropical forest environment typical of other small islands dotted around Hong Kong, replete with birds, butterflies, other insects and flowers, with the Christian signage at each station an added point of interest (see Plate 4.2).

In a nook that was the site of the original chapel constructed by Father Freinademetz, located behind the signposted ancestral home of Vicar General Fr Dominic Chan (now little more than a worn, paved floor with ruined walls partly standing), is a second statue of Father Freinademetz, his gaze directed towards the jumbled stone ruins of his house (also signposted) located about 60 metres away. The two ruined dwellings are linked with a path that is lined with ten large panels depicting in text and photos the life and times of Father Freinademetz so that this space has itself become sacralized.

Another example of the sacralization of a secular site is a 150-year-old village house with no particular religious significance (unlike the ruined house of Father Freinademetz, nor any of the various ruins of houses of the village's Catholic pastors). Just a former family home, this house is now restored and is used as a retreat for silent meditation and prayer. It is signposted as "Prayer Corner", hence sacralized. It is sparsely furnished, it is not set up as a B&B, and although quite large with six rooms, each room is representative of a monk's or nun's cell.

Other examples of the sacralization of secular sites are presented in Chapter 9 on co-creation where, for instance, empty windows in abandoned houses were adorned in 2016 with ceramic and glass mosaic panels portraying scenes from Catholic rites and Hakka customs that graphically combine both the heritage and religious components of 'heri-ligion'.

In the cases of the abandoned village well and the 'Prayer Corner House', heri-ligion exhibits its specific duality, its dialectic nature. In the process of sacralization, these two examples of built heritage have been restored hence heritagized as objects of and for the tourist gaze that is distinct from but linked into their role as symbols of Christianity for pilgrims. They 'speak' to visitors from quite different perspectives, both of which are equally valid in their particular contexts. They attest to Ashworth's (2011, p. 1) cognizance that the interpretation of the past results in *"quite different ways of viewing the past from the present"*. They also represent the fusion between the secular and the sacred in the context of religion as a fundamental constituent of cultural heritage.

References

Ashworth, G. J. (2011). Preservation, Conservation and Heritage: Approaches to the Past in the Present Through the Built Environment. *Asian Anthropology, 10*(1), 1–18. https://doi.org/10.1080/1683478X.2011.10552601

Bai, Y. (2003). *On the Early City and the Beginning of the State in Ancient China*. Bureau of International Cooperation, Hongkong, Macao and Taiwan Academic Affairs Office. Chinese Academy of Social Sciences. http://www.worldlibrary.org/Articles/AncientChineseurbanplanning=Bai,Yunxiang

Chan, D., Cheung, L., & Chan, L. Y. (2012). *The Catholic Nature Trail of Reconciliation*. Caritas Printing Centre.

Chu, W. H. (1998). *Conservation of Terrestrial Biodiversity in Hong Kong* (Unpublished M.Phil. thesis). The University of Hong Kong.

Constable, N. (1994). *Christian Souls and Chinese Spirits. A Hakka Community in Hong Kong*. University of California Press.

Cross, F. L., & Livingstone, E. A. (2005). *The Oxford Dictionary of the Christian Church* (3rd ed.). Oxford University Press.

Graham, B., Ashworth, G. J., & Tunbridge, J. E. (2000). *A Geography of Heritage*. Arnold.

Hall, S. (1997). The Work of Representation. In S. Hall (Ed.), *Representation: Cultural Representations and Signifying Practices* (pp. 13–74). The Open University & Sage.

Knapp, R. G. (1986). *Chinese Landscapes: The Village as Place*. University of Hawaii Press.

Li, F. M. S. (2008). Culture as a Major Determinant in Tourism Development of China. *Current Issues in Tourism, 11*(6), 492–513. https://doi.org/10.1080/13683500802475786

Singh, R. P. B. (2008). The Contestation of Heritage: The Enduring Importance of Religion. In B. Graham & P. Howard (Eds.), *Ashgate Research Companion to Heritage & Identity* (pp. 125–141). Ashgate Publishing.

Sofield, T. H. B., Li, F. M. S., Wong, G. H. Y., & Zhu, J. J. (2017). The Heritage of Chinese Cities as Seen Through the Gaze of Zhonghua Wenhua—('Chinese Common Knowledge'): Guilin as an Exemplar. *Journal of Heritage Tourism, Special Issue on 'Heritage and Cities in China', 12*(3), 227–250. https://doi.org/10.1080/1743873X.2016.1243121

Williams, M., & Webb, R. (1994). Rural Landscapes. In M. Williams (Ed.), *The Green Dragon: Hong Kong's Living Environment* (pp. 113–127). AbeBooks.

CHAPTER 5

Ethnicity and Identity: The Yim Tin Tsai Hakka Heritage Exhibition Centre

Abstract The Heritage Exhibition Centre is a direct manifestation of Hakka ethnicity, traditions and identity that displays many aspects of Hakka culture and plays an important role in the collective memory of the community.

Keywords Hakka ethnicity · Identity · Cultural framing · Unicorn ceremony · Museums · Museology · Symbiosis · 'Heritage on paper' · Collective memory · Entangled anthropology of religion and secularity

Museology (museum studies, museum science) analyses the history of museums and their role in society, as well as the activities they engage in, including curating, preservation, public programming and education (Desvallees & Mairesse, 2009; Murphy, 2018). The transformation of the Ching Po primary school into the Yim Tin Tsai Heritage Exhibition Centre ('folk museum', for short) by the community is also an affirmation of Hakka ethnic and cultural identity, a statement of pride in their specific origins as distinct from the majority Cantonese society surrounding them. It is a statement of their 'otherness' and alterity. A range of studies indicate that the Hakka moved from central and northern China in three major migrations, the first in the early fourth century, the second in the mid-ninth century and the latest in the thirteenth century, with

T. Sofield et al., *Heritage-Making in Hong Kong Through Culture and Religion*, https://doi.org/10.1007/978-981-97-4339-1_5

the majority of them settling in Guangdong (Choon, 2005; Constable, 1994, 2005; Encyclopaedia Britannica, 2013; Erbaugh, 1992; Pletcher, 2013). Globally, there are about 90–100 million Hakka people in total. About 60% of China's 31 million Hakka population live in Guangdong, and about 60 million live in overseas countries (including Taiwan), with about 1.25 million residing in Hong Kong (Lau, 2010). Over centuries their culture, language, cuisine, architecture and social structure began to diverge from the original Han origins, and for about 500–600 years they have been regarded as a distinct linguistic sub-group of Chinese.

The formation of the folk museum progressed through four stages. At the very beginning (before 2005), the concept of turning the former village school into a heritage exhibition centre was the idea of the village representative, Colin Chan, and some village elders. They moved furniture, domestic paraphernalia, memorabilia and photos from their abandoned houses to the former classroom. However, the exhibition room became a mess due to no clear categorization of the exhibits. In the second stage (2005–2006), the villagers set up a heritage exhibition team with a helping hand from Ms. Susanna Siu, Chief Heritage Executive of the Hong Kong Antiquities and Monuments Office, who recruited a group of volunteers from the Catholic Diocese to make interpretation signage for ease of identification. At this point, the Village Committee decided that it needed managerial expertise on the ground and between 2005 and 2007 a couple who were qualified in architecture and design were contracted by the villagers to assist in organizing the museum (stage three). They moved to commercialize the venture as a personal source of income and the arrangement ended in failure because the villagers considered the couple had in effect seized ownership of the building and artefacts in restricting their free access: in village jargon the couple were perceived to have "occupied" their land. This conflict of interest led the villagers to terminate the contract and force them to leave. Subsequently (the fourth stage) the village found Professor Lam Yip Keung, then professor of anthropology and curator of the museum of the Chinese University of Hong Kong and Professor Cheung Siu Woo, professor of social anthropology, Hong Kong University of Science and Technology, to further collect and categorize the exhibits (2007–current date). Professor Lam divided the exhibits into four main categories, namely agriculture, containers (covering a wide range of uses in agriculture, fishing, salt making and storage, family heirlooms, etc.), household items and school life. The first category, agriculture, pays homage to the mainstay of the vast majority of Hakka communities throughout their history, with women always as the main farm workers—one reason advanced to explain the fact that the Hakka women never bound their feet as occurred for several centuries in Han-dominated China, and which contributed to the general image, still widely acknowledged today, that the Hakka women are the hardest workers in Chinese society (Yau, 2016). While salt-making was the foremost industry of Yim Tin Tsai until it ceased in about 1910, rice and vegetable cultivation was always practised on terraced paddy fields on Yim Tin Tsai and the adjoining island of Kau Sai Chau (Atha, 2014) and together with fishing became much more important to the local economy once salt-making ended.

The largest part of the folk museum's collection relates to domestic family life, and the artefacts and their interpretation illustrate many facets of Hakka culture (see Plate 5.1). For example, when a baby girl was born in the village, her family planted a camphor wood tree (*Cinnamomum camphora*) which would form part of her future dowry. This wood is characterized by its pleasant fragrance, is termite proof, the Hakka used its aromatic oils for medicinal purposes and culinary applications, and to ward off clothes' moths. On the daughter's marriage the tree would be harvested and its timber made into furniture, typically a carved camphor wood chest, a small table, stools and a set of shelves (several of which are displayed in the museum) for the new couple. (There is now only one camphor tree, estimated to be 100 years old and heritage-protected, remaining in the village.)

There is one exhibit that merits particular attention as it makes a quite specific statement about Hakka heritage and identity, and that is a display of unicorn head masks, costumes and musical instruments. A significant traditional feature of the folklore/religion of the Hakka was the veneration of the Chinese unicorn (*qilin*), through dance and music. They believed it was an auspicious creature that could bring good luck and ward off evil. It was performed to celebrate the Chinese New Year and occasions like weddings, birthday parties, the inauguration of an ancestral hall, moving into a new home and the birthdays of deities. The dances were also a ritualized form of martial arts since historically, as the Hakka constantly migrated from one region to another in search of new lands, defence was integral to their survival. Each village had its own repertoire of dances and music that functioned to connect the community and unite members of a clan (Hong Kong Intangible Cultural Heritage Office web page, n.d.).

Plate 5.1   School building restored and re-purposed as the Heritage Exhibition Centre with numerous displays

The Hakka use the unicorn dance to differentiate themselves from the Han Chinese who utilize the lion dance for similar ceremonial purposes, and it is thus an important contributor to Hakka identity. It emphasizes their 'otherness' and alterity from the larger Chinese population that envelops Hakka society. While today's older generation of islanders remember that their community used to hold unicorn performances, their last expert in the genre migrated to England "many years ago", and they now have no knowledge of any specific Yim Yin Tsai unicorn dance customs. However, in 2015, a group of UK-resident Hakka from Yim Tin Tsai returned to the village for the official opening of the Exhibition Centre and there performed a unicorn dance, using the equipment that had been left behind in storage in the school when the expert migrated (see Plate 5.2). It was perceived as an appropriate cultural activity to celebrate both the specific occasion and Hakka folklore generally, and as an important community and Chan clan unifier bringing the locals and diaspora together, but it was devoid of any traditional religious meaning. The artefacts remain on display and there has been no move to use them again. In semiotic terms the exhibit is a signifier to both Hakka and non-Hakka as an identity marker. The initiative by the UK diasporic villagers to perform in Yim Tin Tsai is consistent with a worldwide trend where Hakka communities have revived unicorn dances to celebrate the Chinese

Plate 5.2 Hakka traditional Unicorn performance at the opening of the Hakka Heritage Exhibition Centre, 2015

New Year: it represents an important cue for community memory and nostalgia.

The re-purposed village school thus illustrates many and varied aspects of Hakka culture, including their traditional working clothes and formal wedding gowns, the *liangmao* traditional bamboo/straw hat worn by both men and women when farming and fishing (*"the most public symbol associated with the Hakka"* according to Constable, 1994, p. 12) and their distinct cuisine and culinary artefacts. The exhibition combines their ethnic culture with specific examples of their Catholicism, such as a display of a nineteenth-century prayer guide, religious icons and wall hangings, devotional objects, and photographs of past local pastors and church ceremonies, such sacralized items as these 'speaking' symbolically and fundamentally to their current identity as Hakka Catholics. According to the Heritage Exhibition Centre information brochure (Lam, n.d.) all of the artefacts *are direct witnesses and represent the collective community memory of the history and social aspects of Yim Tin Tsai.*

While the collection has been divided into four categories for assembling the many artefacts into a semblance of order that assists visitors to digest them, no such distinctions are made by the villagers. For them each and every artefact has its own personalized history and memory and all are integrated holistically: they encapsulate the three fundamentals of their identity as (i) Hakka, (ii) as Catholics and (iii) as members of the

Chan clan, and each exhibit resounds with multiple meanings in different but harmonious ways.

Not on display in the Exhibition Centre is a significant collection of "heritage on paper" (Professor Lam, personal communications, December 2023), written and printed material found in the abandoned houses that constitute interesting records of the cultural and social lives of local villagers over past decades. These "folk documents" include: (i) *personal records such as school exercise books, paintings, song books, newspaper clippings, family records such as wedding photos, ceremony procedures and records of gifts;* and (ii) *village records such as accounts books of public services covering ferry transportation, local power supply from a generator, and records of church activities. This "heritage on paper" also faithfully reflects villagers' personal and collective lives in the past century* (Lam, n.d.) but is not displayed because of privacy concerns. The collection has been digitized and is now held in the archives of the Hong Kong University of Science and Technology under the custodianship of Professor Cheung, with the approval of the villagers.

While the folk museum exhibits provide insights into Hakka culture for visitors, few lived elements have endured on the island since the village was abandoned (but see Section 8(iii) below on the community's traditional ancestral graves and how they pay respect to their ancestors as a cultural not a religious ceremony). Professor Cheung continues to assist in collecting cultural relics on the island with groups of his students and volunteers from the Yim Tin Tsai community, carefully sifting through detritus in the abandoned houses, recovering many more artefacts, cleaning them and systematically sorting them out into the four main categories for either display or storage. For sustainable development purposes, Prof. Cheung also launched a docent training programme twice a year for his university students and villagers to introduce the heritage of Hakka island life to visitors as museum guides.

The Hakka Exhibition Centre on Yim Tin Tsai, having been established by the source community whose culture it represents, provides an authentic but somewhat truncated overview of Hakka heritage and history on the island since most of the artefacts are less than 150 years old. Its focus therefore captures mainly facets of the lifestyle and living conditions only from the time the community converted to Catholicism in the mid-nineteenth century until the island was abandoned just before the end of the twentieth century. The folk museum not only 'speaks' to visitors about heritage (life and times) of the Hakka community, however:

for former residents it constitutes a form of 'collective memory' that they value highly and to which they can personally relate (see Kirshenblatt-Gimblett, 1998, on observer responses to museum exhibitions). The initiative for establishing it came from them, the work involved in assembling it was largely carried out by them, and the museum thus plays a key role in the resilience of aspects of Hakka culture and traditions.

While the majority of exhibits and artefacts may be considered secular, this small folk museum has embraced spiritual and cosmological perspectives, inevitably reflecting the way in which Catholicism permeated the daily activities and behaviour of the former residents, from the ringing of the church bells at 6 am every morning until they retired at night. For them, photographs of church services, morning prayers in the primary school, weddings and other ceremonies, religious icons and wall hangings, and religious symbols can be used to sacralize the memory of the 'forgotten past', showing the central locus of religion in their former village life. From this perspective, in order to offer an accurate representation of ethnic culture and of the history of Yim Tin Tsai, as lay curators and owners of the material, the villagers consider that their museum would be incomplete without a religious dimension. Through the Exhibition Centre, community memory (collectively and individually) is represented in the assemblage of religious and secular artefacts both as Hakka (cultural identity) and as Catholic (religious identity). It stands as an effective exercise in what Scheyvens and van der Watt (2021) have termed community-based cultural empowerment: "*Cultural empowerment (will) be evident when residents value and respect diverse local heritages; enjoy opportunity and agency to express their cultural heritage in all life domains; and have power to determine how their cultural heritage is portrayed*".

It is also an example of what Isnart and Cerezales (2020, p. 6) call "*converging habituses in heritage conservation that exist in the religious and the secular spheres*". The folk museum they would describe as "*a religious heritage complex*", a term they coined as "*a theoretical tool to capture the co-existence* (in museums and heritage sites) *of two different layers of values attributed to religious practices and materiality*". Meyer and de Witte (2013) argue that the selection of material objects for heritage display involves a form of sacralization: the very framing of such objects, they state, sets them apart, lifts them out of the ordinary, elevates their status as 'true' and authentic representations of a culture and thus imbues them with a sacrality that was absent when a clay urn, for example, was 'just a piece of domestic pottery' in a home.

"Once brought into the framework of heritage, cultural forms are made to assume additional or even new value. The powerful effects of such framing become clear once it is realized that even ordinary everyday objects, coded as heritage, may be elevated to the level of the extraordinary and achieve a new sublime or sacred quality" (Meyer & de Witte, 2013, p. 1)

According to Ksenia Pimenova (2022, p. 13), at the international level:

"Various museums increasingly take into account the voices of source communities and confessional groups. Originally secular institutions, museums have become more permeable to spiritual, cosmological, and confessional perspectives" (and many museums display) "multiple inter-actions, entanglements, and tensions between religion and heritage". (p. 13)

A major issue confronting museum curators in many institutions and at heritage sites is that religious artefacts and symbols are commonly displayed not as part of living culture and lived faith but as works of art or aesthetic representations locked up in glass cases and divorced from their former physical location that was a sacred space or place, even when they are 'authentic' and their origins are accurately interpreted histor-ically, anthropologically and theologically (Sullivan, 2015, on exhibiting sacred religious Asian objects in the secular spaces of European and Amer-ican museums). How to transform their exhibitions to reflect current lived realities is problematic for many curators and Pimenova describes museums as "*new sites for an entangled anthropology of religion and secu-larity ... and these entanglements can be studied from a diachronic (sacred v secular) perspective*" (2022, p. 14). She continues: *In social sciences, the religious and the secular have been analysed as interacting, yet distinct domains,* often acting in opposition to each other. Instead of viewing reli-gion and secularity as opposites, however, we would like to highlight the intertwining and "*elective affinities*" (Luehrmann, 2015, p. 17) between the two that our concept of 'heri-ligion' reveals in our analysis of herita-gization as manifest in Yim Tin Tsai's museum, where symbiosis and its accompanying forms of commensalism and mutualism sow harmony and largely dispel any negative tensions of entanglement.

REFERENCES

Atha, M. (2014). *Archaeological Survey-Cum-Excavation, Yim Tin Tsai, Sai Kung (Oct.–Nov. 2013)*. Hong Kong Archaeological Society.

Choon, Y. N. (2005). *The Hakka Chinese: Their Origin, Folk Songs and Nursery Rhymes*. Poseidon Books.

Constable, N. (1994). *Christian Souls and Chinese Spirits. A Hakka Community in Hong Kong*. University of California Press.

Constable, N. (2005). *Guest People: Hakka Identity in China and Abroad*. University of Washington Press.

Desvallees, A., & Mairesse, F. (Eds.). (2009). *Key Concepts of Museology*. International Council of Museums, Armand Collin.

Encyclopaedia Britannica. (2013, February 26). The Editors of Encyclopaedia. "Hakka". *Encyclopedia Britannica*. https://www.britannica.com/topic/Hakka. Accessed 4 March 2019.

Erbaugh, M. S. (1992). The Secret History of the Hakkas: The Chinese Revolution as a Hakka Enterprise. *The Chinese Quarterly, 132*, 937–968.

Isnart, C., & Cerezales, N. (Eds.). (2020). *The Religious Heritage Complex. Legacy, Conservation and Christianity*. Bloomsbury Academic.

Kirshenblatt-Gimblett, B. (1998). *Destination Culture: Tourism, Museums, and Heritage*. University of California Press.

Lau, Y. C. (2010). *Estimate of Hakka Residents of Hong Kong*. Wikipedia. https://en.wikipedia.org/wiki/Hakka_people

Luehrmann, S. (2015). *Religion in Secular Archives: Soviet Atheism and Historical Knowledge*. Oxford University Press.

Meyer, B., & de Witte, M. (2013). Heritage and the Sacred: Introduction. *Material Religion, 9*(3), 274–280. https://doi.org/10.2752/175183413X13730330868870

Murphy, O. (2018, Spring). *Museum Studies as Critical Praxis: Developing an Active Approach to Teaching, Research and Practice* (Tate Papers, No. 29) [On-line Research Journal].

Pimenova, K. (2022). Museums and Religious Heritage: Post-colonial and Post-socialist Perspectives. *Civilisations Revue Internationale d'Anthropologie et de Sciences Humaines, 71*, 13–28.

Pletcher, K. (2013). *"Hakka"*. Britannica. https://www.britannica.com/topic/Hakka

Scheyvens, R., & van der Watt, H. (2021). Tourism, Empowerment and Sustainable Development: A New Framework for Analysis. *Sustainability, 2021*(13), 12606. https://doi.org/10.3390/su132212606

Sullivan, B. M. (2015). *Sacred Objects in Secular Spaces Exhibiting Asian Religions in Museums*. Bloomsbury Academic, an imprint of Bloomsbury Publishing Plc.

Yau, K. L. (2016). *From Invisible to Visible: Representations and Self-Representaions of Hakka Women In Hong Kong, 1900s–Present* (Master's thesis, Lingnan University, Hong Kong). https://commons.ln.edu.hk/cgi/viewcontent.cgi?article=1008&context=his_etd. Accessed 22 November 2023.

CHAPTER 6

Attachment to Place, Nostalgia and 'Home'

Abstract In this chapter, we explore three components of heri-ligion as mentioned in this chapter's title that assist us to unpack the complexities of the social and cultural processes that have unfolded in Yim Tin Tsai. While we have separated these three elements into separate sections to facilitate understanding, in reality this separation is artificial and attachment to place is multidimensional and intertwined, entangled and connected with nostalgia and 'home' in a variety of ways.

Keywords Ethnicity · Identity · Nostalgia · Attachment to place and home · Symbiosis · Harmony

In considering how to address the evolving situation on Yim Tin Tsai while taking adequate account of the traditional heritage and value system of the Hakka community, we have identified three separate but intimately connected factors, as follows:

  i. The complexities of place attachment;
 ii. Processes underlying a community's drive to define and/or re-discover and re-establish its roots ('home') where religion/religious traditions are of foundational importance, that include specific, pre-existing characteristics of a community (in this case of Hakka

© The Author(s), under exclusive license to Springer Nature Singapore Pte Ltd. 2024
T. Sofield et al., *Heritage-Making in Hong Kong Through Culture and Religion*, https://doi.org/10.1007/978-981-97-4339-1_6

heritage) which may be perceived as major determinants in directing the dynamic pathways of change (ethnicity, cultural value system, geography and so forth); and

iii. Collective imagining and memory reconstruction, expressed as nostalgia.

While we have given subtitles to sections on 'Nostalgia' and 'Home' to structure the chapter, instances of interconnectivity between the components are numerous and conjoin them in multiple ways so that they form an integrated whole in which attachment to place may be seen as all-embracing.

6.1   ATTACHMENT TO PLACE

Attachment to place foregrounds a geographical location that through a range of mechanisms—psychological, social, remembrance of experiences, emotions—bonds a person or community to that place, establishing its role as formative in the life of the individual or community. Those bonds result in differing degrees of 'belonging' to that place. The substantial body of literature on attachment to place reveals that psychology predominates over geography, numerous phenomenological analyses by environmental/human behaviour scholars combine both of the foregoing, and religious studies barely touch on the topic. They provide many definitions weighted to different disciplinary paths, but we are drawn to Scannell and Gifford (2010, p. 1), who reviewed and synthesized a range of definitions into a three-dimensional, person–process–place organizing framework:

(i) *"The person dimension of place attachment refers to its individually or collectively determined meanings;*

(ii) *The process dimension focuses on the psychological and includes the affective, cognitive, and behavioral components of attachment; and*

(iii) *The place dimension emphasizes the place characteristics of attachment, including spatial level, specificity, and the prominence of social or physical elements"* (see Diagram 6.1).

The term 'place attachment' carries connotations of people bonding to the geo-physical environment, but the model produced by Scannell and

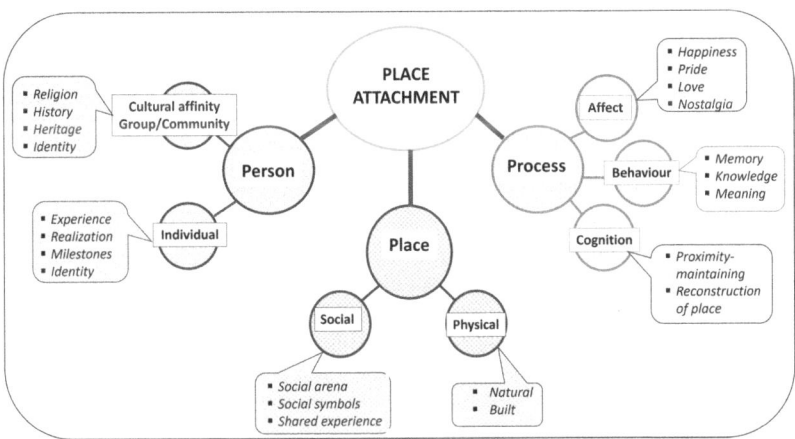

Diagram 6.1 Tripartite model of place attachment

Gifford (2010) demonstrates that attachment may be formed and developed equally through social relationships and/or psychologically-derived responses. Meanings, feelings and experiences associated with that place may be as important to the attachment process as the physical entities of a landscape (Riley, 2012). More precisely, places are: *"repositories and contexts within which interpersonal, community and cultural relationships occur, and it is to those social relationships, not just to place qua place, to which people are attached"* (Low & Altman, 2012, p. 6).

The term, 'topophilia', made popular by human geographer, Yi-Fu Tuan (1974) is analogous to place attachment and often used interchangeably. Taken from the Greek *topos* meaning 'place'; and *philia* meaning 'love of' (Oxford English Dictionary, 2020). Tuan's topophilia is defined as a strong sense of love for place that i*ncludes all of "the human being's affective ties with the material environment"* that shows great variations from person to person (p. 93). His definition has been critiqued as being somewhat restricted in application because of a reflexive methodology emphasizing humanism, existentialism and universalism of cultural traits (Hubbard & Valentine, 2004), and is not as broad as the approach adopted by Scannell and Gifford (2010), Low and Altman (2012) and others. It thus limits a wider intradisciplinary appreciation of the multi-dimensional character of attachment to place. Place attachment is an integrating concept that contributes to individual, group and cultural self-definition and integrity (Scannell & Gifford, 2012), and as Ashworth and Graham (2016, p. 1) so accurately note: "*senses of place interact with senses of ethnic/cultural identity*", while "*government, media, residents and tourists* (all play differing roles) *in creating senses of place, which change through time*". Expanding on this, Giuliani (2003) asserts that:

> There is perhaps no feeling of mutual affinity, community, fraternity among persons, whether formal or informal, institutionalized or not – nor feeling of diversity, aversion, hostility - that is not in some way related to matters of place, territory and attachment to places. This has far-reaching implications. The feeling we experience towards certain places and to the communities that the places help to define and that are themselves defined by the places - home (family, relations, friends), workplace (colleagues), church (fellow worshippers), neighbourhood (neighbours), city, country, continent – has a strong positive effect in defining our identity, infilling our life with meaning, in enriching it with values, goals and significance.

For the Hakka ethnic community of Yim Tin Tsai all of these variables intermix to produce a range of heritage-related narratives that result in differing degrees of 'belonging' to the island. Affective bonds arise from those emotional responses to events, incidents and encounters experienced within the boundaries of a particular place and such emotional qualities are often accompanied by cognition (thought, knowledge, belief) and practice (action and behaviour) (Low & Altman, 2012, p. 5). That is, place attachment involves an interplay of affect and emotions, knowledge and beliefs, behaviour and actions in reference to a geographical place (Proshansky et al., 1983, cited in Low and Altman, 2012)—a tripartite composite of person–place–process as delineated by Scannell and Gifford (2010) (see Diagram 6.1).

Nomenclatures (naming) is intrinsic to community identity, not only as being necessary for defining one's roots but also for claims of territorial ownership. As Reid (2016) states: "*The process of naming is more than a value-free description of a point in space, being a means of expressing and fostering senses of place and linking these with selected aspects of the past*", and the very naming by the Hakka first settlers 370 years ago of their new place of abode as Yim Tin Tsai was in memory of their original village in Yantian District in Guangdong. Naming, however, is not only an official or organizational function: at the personal level, individuals will confer special labels and 'nicknames' on their own favoured, possibly secret places (a hidden cave, a secluded cove, a hollow inside a tree, a strategically-positioned boulder, etc.), many of these labels arising from childhood games. These highly personalized sobriquets may be a trigger for arousing a profound sense of attachment to place, born of nostalgia, that an individual may carry as a marker of identity throughout their life no matter how far they wander from that geographical heart. The stories of youthful adventures and events on Yim Tin Tsai as recounted by former residents and recorded through the Community-Based Narratives research project attest to the power of names in revealing attachment to place. The freedom the villagers had to name their island and personal places constitute examples of both community and personal empowerment.

In the first instance, Hakka folklore meant that Yim Tin Tsai was interpreted from a cultural perspective where *feng shui* played an essential role in the selection of a location for their new settlement, consequentially bestowing names that had survived since antiquity on surrounding geographical and physical features. But over time, as the community transitioned to Catholicism, those traditional principles lost their significance and were replaced by their adopted religion, most profoundly in the siting of their chapel on top of Black Tortoise Ridge and the construction of the Nature Trail as a Christian journey of discovery for pilgrims and others. This transformation constitutes an example of heri-ligion in a novel guise where the heritage of one's original religion and world view has been replaced by a very different religion and world view. However, the persistence of some traditions from Hakka folklore combined with the practices of contemporary Catholicism (syncretism) are both intrinsic to individual and community identity. When the Government of Hong Kong designated St Joseph's as a heritage-protected building, and UNESCO subsequently awarded its Certificate of Merit for Culture Heritage Conservation, the character of the building was significantly altered, with consequences for sense of place and attachment. From simply a prominent building (if the most important one) in the village which was more or less unknown to people outside the wider Catholic community of Hong Kong, the church was elevated to a monumental and historic status to which, as heritage, outsiders could relate. It was no longer just a minor building in an abandoned village but was reframed—and in effect therefore re-named—as an important Heritage Monument with its dual official accreditations by the HK Government and UNESCO.

The issue of scale then becomes relevant: In what context is an attachment to Yim Tin Tsai exhibited by larger entities? Or to larger geographical spaces? For example, when we add to the mix of heritagization on Yim Tin Tsai the interpretation of the wider Hong Kong Catholic community—and indeed the global Catholic community—a much large scale is generated. The Vatican's elevation of Father Freinademetz to sainthood may create a sense of 'proprietorship' for Catholics internationally and not just in Hong Kong that has nothing to do with historical personal association and ancestral roots. If we then include UNESCO's awards recognizing the heritage significance of the restoration of the chapel and the salt fields for appreciation by any traveller visiting Yim Tin Tsai, we observe that the creation and imagining of place is thus *"user-determined, polysyllabic and changeable over time"* (Ashworth & Graham, 2016, p. 1), and the folk museum plays a significant role in this process. Since community members and even whole cultures often consensually or collectively share attachments to places, a variety of collective group or cultural place attachments may transcend the unique experience of individuals (Low & Altman, 2012). This then raises the question of whether attachment to place is affective, derived from personal experience(s) or whether it is symbolic in that representations of what Yim Tin Tsai signifies may engender some sort of attachment? It is considered useful, therefore, to differentiate between affective and symbolically-oriented attachment. With reference to Yim Tin Tsai, for members of the Chan clan who once lived there the primary attachment to place may be affective, operational at both the individual and collective levels. For younger generations who have never known the place as their physical home it may still have remnants of individual affective responses derived from their parents' lived experiences but be more symbolic because of their physical and emotional distance from the village. And for the broader Catholic community of Hong Kong, their primary relationship is more doctrinally communal and centred in the realm of the symbolic, although interviews with some of those who are/have been active volunteers and have participated directly for a decade or more in bringing the village indicate that they have developed a personal affective relationship with the place. We may therefore distinguish attachment to Yim Tin Tsai as a place from three perspectives and scales: (i) for the former inhabitants it constitutes an 'ordinary' landscape associated with the routine of everyday life as formerly enacted; (ii) for the Yim Tin Tsai diaspora, especially the younger generations who have never lived there, it signifies their ancestral 'home', a landscape replete with symbolic values of clan and family histories; it may evoke 'romantic' notions of home not founded on personal experiences. And: (iii) for the Catholic community of Hong Kong it represents a 'special' landscape as signified by the shrine to Father Freinademetz, St Joseph's Chapel and associated religious connotations, in contrast to the attachment held by the former inhabitants.

For Yim Tin Tsai, we may conclude that intensity of attachment to place varies among individuals and at scale but that, overall, attachment to place by former residents has been a key driver in the various endeavours to bring Yim Tin Tsai 'back to life' since. This is because, along with the idea that attachment is a positive, affective link between an individual and a specific place, one of its central features is the desire to maintain close relations with that place (Hernández et al., 2007). In other words, despite increasing mobilities and globalization processes, place continues to be an object of strong attachment (Lewicka, 2011).

6.2   NOSTALGIA

Nostalgia is a recollection of the past, often characterized in personal reflections as a period with happy associations covering personalities, places and events (Boym, 2002). Important functions of nostalgia include *inter alia* social connectedness, memory consolidation and existential meaning (Wildschut et al., 2006). It typically invokes memories of people one was close to, such as family members, neighbours, childhood friends, and others and thus increases a sense of social support and connections. Emotions and feelings are central to the concept. Thus, feelings about women as wives/mothers and the home in which they nurtured their family (Ahrentzen, 2012), childhood memories of a loved place, and adult feelings towards the place where they made their home (Marcus, 2012), will reinforce *"emotional embeddedness, feelings of security, esteem and belongings associated with places"* (Brown & Perkins, 2012). These affective factors are particularly manifest in family gatherings and community events such as church services when diasporic members return to Hong Kong: such occasions invariably produce shared reminiscences of times past growing up on Yim Tin Tsai. The numerous stories, assembled by the Community-Based Narratives research team (Sham et al., forthcoming) whose work is ongoing, provide numerous additional emic insights into Hakka culture and identity, and the first fifty of these stories may be accessed on the website https://www.yttstorytelling.org/en/storymap.

They are replete with nostalgia, which acts like an unsevered umbilical chord linking all of the villagers—whether part of the Yim Tin Ysai Hakka diaspora living overseas in Canada or England or Australia or residing locally elsewhere in Hong Kong—to the island with invariably happy memories of their former life there. Sedikides and Wildschut (2018, p. 48) define nostalgia as a sentimental longing for one's past that is at once both personal and social: "*It does so primarily by increasing social connectedness (a sense of belongingness and acceptance), and secondarily by augmenting self-continuity (a sense of connection between one's past and one's present)*". Nostalgia, according to Routledge et al. (2012, p. 452), "*serves as a motivator for the preservation of a people's cultural heritage* (and) *the power of the past* (elevates) *nostalgia as a meaning-making resource*". In other words, nostalgia for past times may function as a stimulus for the preservation of one's cultural heritage through conservation of buildings, landscapes, and other artefacts of historical significance (Tanselle, 1998): and the restoration of St Jospeh's church and the establishment of the Heritage Exhibition Centre are concrete examples of this. Tanselle (1998) would interpret these endeavours as clearly motivated by a collective nostalgic determination to connect their heritage to past generations. The annual visit to ancestral graves on Kau Sai Chau may also be construed as one manifestation of this tendency (see the following section 8(iii) for details). The major annual church services such as St Joseph's Feast Day and All Saints Day which have been continuously celebrated on Yim Tin Tsai for more than 150 years (even after the village was abandoned), and which attract overflowing congregations, are additional manifestations of enacting living history. Nostalgia as a key underlying force thus further facilitates community socialization and cohesion (Wildschut et al., 2006). The Yim Tin Tsai stories of its former residents represent heritagization through nostalgia and memory and hence contribute to overcoming the loss of traditions through modernization.

An artistic manifestation of Yim Tin Tsai nostalgia occurred in 2019 through a series of five glass paintings collectively titled 'Nostalgia' that (as was the case in 2016 when Hulu Culture artists carried out similar works of art) were located in the empty window frames of abandoned houses. The artist, Candice Keung Lap-yu, drew inspiration from "the unforgettable memories" of five villagers about their former life on the island. They covered a combination of religious and secular scenes such as a Hakka woman working on the land at sunrise; traditional wedding scenes such as the "crying marriage" (when a bride was taken from her parents' house to marry and live with the groom's parents, the mother would not go to the ceremony but stay at home and cry for three days to express her sadness at the loss of a daughter and a major contributor to the family's welfare); and Catholic church scenes such as a baptism and an eucharist service. The paintings were part of the Hong Kong Tourism Commission's sponsorship of the Yim Tin Tsai Arts Festival (see Chapter 9 below on co-creation for more detail). The material of the paintings (glass) and their placement in windows simulated stained-glass windows in churches so that an element of sacralization enveloped each painting regardless of subject matter. In the context of our concept of heri-ligion we would assign this artistic expression of nostalgia to a seamless mix of the sacred and profane, a symbiotic relationship that also includes elements of syncretism where traditional marriage customs sit comfortably with Catholic marriage sacraments. As an affective response to growing up on Yim Tin Tsai, it shares obvious characteristics with, and may be seen as making a direct contribution to, attachment to place. It is of interest that the root of the word 'nostalgia' is the Homeric Greek term '*nostos*' meaning 'homecoming' (Fuentenebro de Diego & Ots, 2014) so there is also a direct link into our following section on 'Attachment to Home'.

6.3   ATTACHMENT TO 'HOME'

For Chinese, whether residing overseas or whether they now live in mega cities such as Beijing, Shanghai, Guangzhou or Hong Kong, 'home' will often be described in terms of 'roots', i.e. their ancestral village. Even when they have not been born in that village, the concept of ancestral roots remains strong, and part of the self-identity of many Chinese who will relate strongly to that place of origin (Maruyama et al., 2010). 'Roots' provide a direct linkage to ancestors, and this phenomenon is played out in the lives of the Yim Tin Tsai community. For Chinese emigrants who journeyed overseas, for example to gold rushes in the United States, Canada, South Africa, Australia and elsewhere in the nineteenth century, the concept of 'sojourner' was conceived to describe their continuing spiritual attachment to their 'home' village and where they expected to return one day or, if they died overseas, to where their remains would be sent for burial (Sinn, 1989). As sojourners, the 'home' they established overseas was often reduced to the simple necessity of having a roof over their head and not permanent: 'roots' were more important than a 'roof' (Somerville, 1992). This desire was not confined to the nineteenth and early twentieth centuries. For example, when Mao Zedong and the Communist Party gained control over China in 1949 and closed the 'Bamboo Wall' to both the living and the dead, the desire of many overseas Chinese to join their ancestors in their village grave site was thwarted. In response, a prominent Hong Kong *taipan* and philanthropist, Lee Iu Cheung, in 1950 established a columbarium to receive and hold the ashes/bones of overseas Chinese free of charge until such time as China might relax its ban and it would be possible to return them to their ancestral homes. The columbarium is still extant today, run by the Li Family Trust, to provide a service for poor families unable to afford the costs involved (Dr. Lee Cheung, son of Lee Iu Cheung, personal correspondence, 2017).

For centuries a strong sense of identification with one's village has been entrenched in the location of family grave sites just outside the residential section of the village, in the vicinity of their home and kin, and this is common for thousands of villages all over China and Hong Kong. The importance of being buried in the ancestral grave site is fundamental to the identity of millions of Chinese and the annual '*ching ming*' ('grave cleaning') festival which takes place in early Spring (usually April) will bring descendants together, often including those from overseas, especially but not only where the Daoist veneration for ancestors remains potent. For millions of Chinese, participating in the *ching ming* is seen as virtually mandatory and it witnesses one of the great mass mobilities in contemporary Chinese society. A second ceremony which includes veneration of ancestors is the *chung yeung* ('double nine') Festival when families also visit the graves of their ancestors to pay their respects. According to the ancient '*I Ching*' ('*Book of Changes*'), one of the oldest of the Chinese classics dating back more than 2500 years, the ninth day of the ninth month in the Chinese calendar has very powerful '*yang*' and is thus one of the most auspicious days of the year. Traditionally, Chinese would climb a high mountain, drink chrysanthemum tea or wine, and hold a feast; and as a general holiday in Hong Kong today these practices are still observed by many. Extended family groups will also visit their ancestors' graves in the hills around the New Territories and replicate many of the practices of the *Ching Ming* festival such as cleaning the grave sites, repainting inscriptions, and offering roast suckling pig and fruit to be symbolically consumed by the ancestral spirits during prayers, and then eaten as a picnic feast.

For Christian Chinese, some will not participate in the *ching ming* and *chung yeung* festivals, but others will do so out of general respect for ancestors and rationalize them as cultural events, perceiving their actions as joining in a traditional Chinese ceremony rather than a religious ritual. This is particularly so for Catholic Chinese because the Catholic church accepts a degree of *"incorporation of Chinese forms and symbols"* into Catholic rituals (Constable, 1996, p. 18). There is thus no contradiction between being Catholic and Chinese as expressed through acceptance of many practices, particularly those that are *"related to death and ancestors* (which) *are transformed or rationalized to reconcile them with what is considered pious Christian belief and practice"* (Constable, 1996, p. 19). In the context of heri-ligion, this constitutes an example of both syncretism and commensalism.

For the Hakka of Yim Tin Tsai their Chinese identity and attachment to place is located in their common origin, genealogies and history (Chang & Lin, 2022; Watson, 1988) and their ancestral graves are intrinsic to who they are. They still acknowledge Yantian in Guangdong as their 'original' home historically, and prior to becoming Catholics the Yin Tin Tsai community buried its dead in graves on a site that captured 'good' *feng shui* on the nearby island of Kau Sai Chau. But in leaving Daoist ancestral worship more than 150 years ago and transferring their cemetery to Yim Tin Tsai, burying their dead as Catholics has reinforced their changed status as Christians, divorced them from past 'pagan' practices and effectively transplanted their roots to Yim Tin Tsai. Shek (foundation Director of "Salt & Light") advised that the Yim Tin Tsai Village Committee does not organize any celebration for either the *ching ming* or *chung yeung* festivals, but with the Catholic Church the community celebrates All Souls Day in November each year, which the villagers call *"the day to worship ancestors"*. The Committee organizes a trip to visit the ancestral graves in both Kau Sai Chau and Yim Tin Tsai in the first week of November. In 2021, Shek (personal correspondence, February 2022) described it as *"a unique experience* (mixing) *Hakka culture and Catholic religion"*. First the group honoured the Chans' ancestors in Kau Sai Chau following Daoist traditions but as a cultural not religious event; and then they travelled back to Yim Tin Tsai and honoured their ancestors in the Catholic cemetery, observing Catholic religious traditions, before finally attending a mass in St Joseph's Chapel presided over by Father Dominic Chan Chi-Ming. Prior to COVID-19 the village committee would book tables in seafood restaurants in Sai Kung to revere elders

Plate 6.1 Hakka traditional ancestral graves on Kau Sai Chau and contemporary Catholic cemetery on Yim Tin Tsai

as well, but it was cancelled in 2022 and instead, after the mass, the gathering enjoyed a roast suckling pig feast on Yim Tin Tsai (see Plate 6.1).

The community's Catholic grave site on Yim Yin Tsai is located about 300 m from the village on the back of a northwestern hill out of sight of the village. After the village was abandoned, the path to the graves was quickly overgrown in the sub-tropical climate and access became difficult. One of the current benefits mentioned by several former residents is that the 'Catholic Nature Trail of Reconciliation' includes a spur that runs off the main path for about 200 metres through the jungle to the graveyard, and because of constant maintenance of the Trail it is now a straightforward stroll without the need to clear the overgrowth and disturb snakes along the way! Yim Tin Tsai's celebration of All Souls' Day to honour their ancestors as a mixed religious and cultural festival now situated within the religious context of Catholicism bears similarities with the HERA studies in Europe where once-sacred rites have been heritagized (HERA, 2020). They include the Dutch project which examined the performances of Passion plays during the Dutch Easter celebrations as an instance of using the religious past in shaping a present-day secular-but-religiously-based Dutch national identity (HERA, 2020). In the case of Yim Tin Tsai the original religious base has been transformed from ancestor worship to a Christian event where Christianity has been the catalyst to convert the Daoist component of the day's events into a cultural observance for the community. The Yim Tin Tsai situation may also be viewed from the perspective of lived Catholicism in the context of the heritagization and festivalization of a millennia-old Chinese tradition, with some similarity to a component of the Polish HERA project that *inter alia* studied examples of intangible heritage associated with Catholicism within Krakow's cityscapes (HERA, 2020). On Yim Tin Tsai we thus have another example of syncretism and mutualism.

This ease of access to the Catholic graveyard, which as mentioned is the sixth 'station' of the Nature Trail that symbolically represents resurrection, results in many secular tourists to the island also viewing the tombstones throughout the year as an attraction or point of interest but with little or no personal religious affiliation or attachment, a somewhat different example of the secular heritagization of a sacred site in the context of heri-ligion-initiated visitation. For non-Christian Chinese, the graveyard is still perceived as a-sacred-site-now-heritagized since it retains its sense of religiosity because of the characteristic Chinese traditional respect for ancestors. Non-Chinese secular visitors also tend to respect the site as a sacred place while following none of its specific Christian origins. For many of them, a key point of interest is a distinctive Portuguese altar erected at the cemetery about 100 years ago as a gift to the Catholic community of Yim Tin Tsai by a departing Portuguese Army unit from Macau.

Another perspective on respect for ancestors, attachment to place and home, and sacralization arose from the enthusiastic participation of many villagers as volunteers in the restoration of the salt pans. Professor Chueng (personal communications) perceived *"three integrated elements of spirituality in their motivation that were palpable and profound"*. Their commitment and endeavours connected them spiritually in three ways:

(i) to their original home of Yim Tin in Yantian, Guangzhou, and to their Chan lineage and to their Chan ancestors who made their livelihood from salt-making spanning centuries until they ceased production in 1910 in their former Yin Tin Tsai home. Contextually, for them this was a cultural embrace embedded in the Hakka tradition of attachment to home and respect for elders and not to be confused with ancestral worship that might cut across their Catholicism in any way;

(ii) to their Catholicism through the cardinal significance of salt biblically. They are familiar with Jesus' Sermon on the Mount where he referred to his disciples as: 'The salt of the earth' and 'the light of the world' (Mathew ch.5: verses 13–14). They will hold a crystal of salt up to the sun and declare: "Look! There is a cross inside!", linking them contextually to their faith. Additionally they have fashioned a series of biblical miniature souvenirs such as crosses and angels made from salt; and

(iii) to Nature, to be good stewards of the environment as encouraged by the Bible. Working to restore the salt fields was reaching out to commune with creation. This was a labour of love, perceived as biblical love, devoid of commercial or mercenary intentions, with nostalgia for their past intimately linked to their current worldview determined to a significant extent by their Catholicism.

The multiple spiritualities that the villagers derive from the restoration of the salt pans confer a strong degree of sacralization on them, and the result is an example of the sacred/secular gaze where outsiders see little or nothing of the biblical connotations that the villagers see: it encapsulates in microcosm our concept of heri-ligion.

UNESCO's 2015 Industrial Heritage Award for the restoration of the salt fields, lacking any reference to religion or spirituality, was announced publicly at a special church service held for the occasion in St Joseph's Chapel, reinforcing the community's ambient understanding of the spirituality and sacralization of the salt fields. This sense of spirituality was not shared by other workers on the salt fields restoration nor UNESCO, nor secular tourists and we have here another example of Ashworth and Graham's (2016) polysyllabic interpretation of heritage.

In considering issues of identification related to place and heritage, the question of scale becomes significant. Every place exists in a hierarchy of spatial scales—local, regional, national and global, and as Graham et al. (2000, p. 2) note: *this basic geographical characteristic has implications for identity and identification* (because) *places at different scales may have different identities which may be mutually reinforcing, irrelevant or contradictory.* The meaning (s) ascribed to heritage extant at each scale may hold distinctive attributes for the identity of individuals, communities and nations. Thus place, in the sense of spatial scale, becomes a key variable in identity in addition to the socio-cultural elements such as Chinese-ness, Hakka heritage, language and religion that these attributes invoke. Yim Tin Tsai captures this multiplicity of meanings and sense of attachment in a descending order of intensity:

(i) At the local scale, for former residents (especially older villagers) the sense of attachment is strong and the combination of Hakka heritage, shared lived experience and Catholicism has elevated the church to a primary role in their individual and collective identity, with the transformation of the school into the Exhibition Centre of "2000 years of Hakka Culture", the restoration of the salt pans and re-purposed village houses reinforcing the individual and community dimensions of place attachment;

(ii) For former residents of Yim Tin Tsai now living overseas (the Yim Tin Tsai diaspora, signifying a 'stretched' local scale) attachment to the island as 'home' is decreased by the commitment to forging life in their new 'home' (as expounded above) but still important to their identity for many of them;

(iii) The wider Catholic community of Hong Kong has enthusiastically adopted Yim Tin Tsai's religious heritage based on the historical residency of Father Freinademetz because the island's unique status grants by association to them a degree of local and global recognition;

(iv) At the scale of the broader Hong Kong entity there is pride in the UNESCO heritage awards granted to Yim Tin Tsai, reinforced by official Government acknowledgement of Yim Tin Tsai in its Tourism Commission's promotions and

(v) At the global scale, Catholics everywhere who may be acquainted with the island through the papal endorsement of Father Freinademetz as a saint, may feel a spiritual connection to Yim Tin Tsai, albeit somewhat abstract and distant.

While the intensity of attachment diminishes as scale increases, each of the scales engenders responses that are mutually reinforcing, with heriligion as a key factor in the situation.

Scale is only one aspect of the complexity that is inherent in any consideration of place attachment. Easthope (2009, p. 62) asks the question: *"Is increasing mobility leading to increasingly dislocated identities?"* He posits that *"the spatial, social, and temporal movement of people, goods, money, and ideas* (all affect) *the nature of identity construction"* in different ways and to different extents. In the context of geographical discourse of identification with place, with reference to Yim Tin Tsai it is reasonable to suggest that the mobility of its inhabitants has led to increasingly dislocated identities, as evidenced by Colin Chan's struggles as he strived to reconstruct the village. The relocation of so many of the original village inhabitants away from Hong Kong to diverse countries, specifically among the younger generation who have been born overseas and no longer inculcated in the lore of the village and the Chan clan where the Hakka social environment enveloped them and dominated most aspects of life, has resulted in hybrid and flexible forms of identity that are inevitably replacing identity based on attachment to place. Studies of second or third generations with longer migration histories (e.g. Leo, 2013 on the Hakka diaspora) reveal that 'home' is conceptualized at a more generic or national level, which may or may not include the specific community of origin. They lack the personalized roots of association with the birth village of their parents.

The lack of enthusiasm to 'return home' to the village that Colin Chan encountered may also be explained via a relatively straightforward economic rationalization: Yim Tin Tsai offers no prospects for the type of employment and income generation that they currently enjoy in Hong Kong or overseas. They may retain a degree of emotional attachment to their ancestral village but that is insufficient to motivate them to re-settle there. The situation presents the two sides of an entanglement coin that posits villagers in divergent groups: resistance to reconstruc-tion of the village based on rational economic reasoning on one side, and an emotional desire to retain and restore one's ancestral home that encapsulates 370 years of families' histories on the other.

On another level, in terms of general studies on migrants, it can be argued that in the twenty-first century 'home' for migrants is increasingly a contested concept, set against three main themes which define contrary pressures that militate against returning. Wong and Wang (2007, p. 170) identify the first theme as asserting that:

> Home is completely impossible for some migrants. Whenever people have left their first and original social milieu [to migrate], they will be unable to come back home again and are doomed to be left in a state of 'impossible homecoming' because, [according to Chambers, 1994, p .1], with the acceleration of globalization, migrants are 'increasingly confronted with an extensive cultural and historical diversity that proves impermeable to the explanations we habitually employ about 'home'.

The second theme identified by Wong and Wang (2007, p. 170) suggests that even though 'going home' is *"not completely impossible, it can only be a 'broken mirror'*", that is, it can only be an *'imaginary homeland'* limited to the mind that is not the reality. In the same vein,

Rushdie (1991, p. 10) argues that for migrants, *'home is not restorable for any geographical location, but exists in the hearts of migrants'*. Similarly, Harman (1988, p. 89) suggests that memories or feelings of 'home' are *'like a moveable feast that migrants could 'bring along like a suitcase' and set up wherever they decamp.*

The third theme identified by Wong and Wang (2007) draws on the claim by Said (1984, p. 170) that for migrants home is undesirable:

"In a secular and contingent world, homes are always provisional. Borders and barriers, which enclose us within the safety of familiar territory, can also become prisons, and are often defended beyond reason or necessity'. Therefore, only by being released from home can migrants become truly liberal and strong."

These three propositions may be applied in varying degrees to some former residents of Yim Tin Tsai, where nostalgia about the 'good things' is subservient to lived experiences of a decaying village enmeshed in chronic poverty, and it cannot compare with the material wealth and standards (housing, income levels, education, health, etc.) that they currently enjoy in their new homes in Canada, England, Australia or mainland Hong Kong. And for the children of original Yim Tin Tsai inhabitants who have only ever experienced life in their parents' 'new' home, nostalgia (if it exists) for the island village will be an imaginary construct, even if they and their parents return for occasional visits and reminisce to them about what it was like to live there.

Although we have not researched this specific point, we consider it reasonable to also assume that the monolithic adherence to Catholicism practised in the village has dissipated to some extent as younger generations, immersed in the western culture of their adopted countries, begin to mirror the increasing secularization and atheistic trend away from all religions (see, e.g., Bauman, 2001 on "The Individualized Society"), and this will be a further point in diffusing Yim Tin Tsai's ancestrally-grounded connectivity. As the longevity of separation from Yim Tin Tsai increases, indigenous language skills diminish and the personal relationships held by the older generation disappear, we anticipate that the annual pilgrimages by such families will also lessen significantly. This suggestion is consistent with the findings that some migrants *"remain rooted in their home place and do not change, others integrate fully with the new host society, some develop multiple place attachments, remain simultaneously mobile and rooted, and some become rootless"* (Gustafson, 2001; McHugh & Mings, 1996; cited in Li & McKercher, 2016, p. 362). Leo (2015), whose study focused on 'global Hakka' living in different parts of the world, found that the relative isolation of many of them meant that *"such families had to adapt radically to their new environment and compromise on their Hakka identity … with constant shaping and reshaping of Hakka identity* (in) *the twenty-first century in the context of globalization, transnationalism, and deterritorialism"*. However, while Hakka identity may be elastic for many who have lived overseas for one or more generations, for those living in Hong Kong a wide range of cultural markers such as language, customs, behaviour, cuisine, gender roles, attire, architecture, and others, retain their validity. For former Hakka residents of Yim Tin Tsai who reside in Hong Kong, the original attachment to place and identity remains vibrant, reinforced by:

(i) Continued strength of Hakka heritage—two active Hakka associations with thousands of members are maintained in Hong Kong;

(ii) Catholicism in the Special Administrative Region of Hong Kong where the Vicar General's links to Yim Tin Tsai are known to all Catholic believers; and

(iii) Ease of access to Yim Tin Tsai by HK residents where multiple visits throughout the year are possible, in stark contrast to many of those who live overseas.

One of the exhibits in the 2021 Yim Tin Tsai Arts Festival sought to encapsulate the multitude of sentiments, emotions and connection to home in a sculpture entitled '*The House of Longing*'. In the words of the artist:

"In the 1960s and 70s, many villagers had to move away from Yim Tin Tsai to look for jobs, and the humming island fell into disrepair along with the houses on it. A lot of these émigrés missed their homeland badly and longed for the terroir as well as the people. These village houses are one of their memory carriers. The sculpture is inspired by the ancestors of the Chan clan, pioneering settlers who went forth and multiplied. Each house also represents the sentiment of interlinked families in an extended genealogy. The affection perseveres, and the longing endures to the end."

Plate 6.2 'Home'

(www.YimTinTsaiArtsFestival2021.hk accessed 10 August 2022, see Plate 6.2)

Colin Chan summarized the attachment of the Yim Tin Tsai community to ancestors, home and roots in an interview in March 2021: "*In our hearts, we villagers respect our ancestors. We respect our elders. Even though I went abroad for school, whenever I come back, I feel a sense of family and belonging. This is the land of our roots. You can't change that*" (Huang, 2021).

REFERENCES

Ahrentzen, S. B. (2012). Home as a Workplace in the Lives of Women. In S. M. Low & I. Altman (Eds.), *Place Attachment: A Conceptual Inquiry* (Ch. 6, pp. 113–123). New York and London: Plenum Press.

Bauman, Z. (2001). *The Individualized Society*. Polity Press.

Boym, S. (2002). *The Future of Nostalgia*. Basic Books.

Brown, B., & Perkins, D. (2012). Disruptions in Place Attachment. In S. M. Low & I. Altman (Eds.), *Place Attachment: A Conceptual Inquiry* (ch.13, pp. 279–299). New York and London: Plenum Press.

Chang, C.-C., & Lin, Y.-H. (2022). Constructing Hakka Ethnic Identity Through Narrative Genealogy Writing. *SAGE Open, 12*(1), 215824402210799-. https://doi.org/10.1177/21582440221079913

Easthope, H. (2009). Fixed Identities in a Mobile World? The Relationship Between Mobility, Place and Identity. *Identities: Global Studies in Power and Culture, 16*(1), 61–82.

Fuentenebro de Diego, F. & Valiente Ots, C. (2014). Nostalgia: A Conceptual History. *History of Psychiatry, 25*(4), 404–411. https://doi.org/10.1177/0957154X14545290

Giuliani, M. V. (2003). Theory of Attachment and Place Attachment. In M. Bonnes, T. Lee, & M. Bonaiuto (Eds.), *Psychological Theories for Environmental Issues* (pp. 137–170). Ashgate.

Graham, B., Ashworth, G. J., & Tunbridge, J. E. (2000). *A Geography of Heritage*. Arnold.

HERA (Humanities in the European Research Area). (2020). *Hereligion. Heritagization of Religion and Sacralization of Heritage in Contemporary Europe.* http://heritagization.eu/ebook/

Hernández, B., Hidalgo, C., Salazar-Laplace, M., & Hess, S. (2007). Place Attachment and Place Identity. *Journal of Environmental Psychology, 27*(4), 310–319.

Huang, G. (202,1 March 3). *Yim Tin Tsai: Hong Kong's Last Salt Village*. Goldthread. https://www.goldthread2.com/travel/hong-kong-yim-tin-tsai-salt-flat/article/3126605

Leo, J. (2015). *Global Hakka: Hakka Identity in the Remaking*. Brill.

Lewicka, M. (2011). Place attachment: How Far Have We Come in the Last 40 Years? *Journal of Environmental Psychology, 31*(3), 207–230.

Li, T. E., & McKercher, B. (2016). Effects of Place Attachment on Home Return Travel: A Spatial Perspective. *Tourism Geographies, 18*(4), 359–376. https://doi.org/10.1080/14616688.2016.1196238

Low, S. M., & Altman, I. (Eds.). (2012). *Place Attachment: A Conceptual Inquiry*. Plenum Press.

Marcus, C. C. (2012). Environmental Memories. In S. M. Low & I. Altman (Eds.), *Place Attachment: A Conceptual Inquiry* (Ch.5, pp. 87–112). Plenum Press.

Maruyama, N. U., Weber, I., & Stronza, A. L. (2010). Negotiating Identity: Experiences of "Visiting Home" Among Chinese Americans. *Tourism, Culture & Communication, 10*(1), 1–14. https://doi.org/10.3727/109830410X12629765735551

Oxford English Dictionary. (2020). Oxford University Press.

Proshansky, H. M., Fabian, A. K., & Kaminoff, R. (1983). Place-Identity: Physical World Socialization of the Self. *Journal of Environmental Psychology, 3*(1), 57–83.

Riley, R. B. (2012). Attachment to the Ordinary Landscape. In S. M. Low & I. Altman (Eds.), *Place Attachment: A Conceptual Inquiry* (ch.2, pp. 13–35). Plenum Press.

Routledge, C., Wildschut, T., Sedikides, C., Juhl, J., & Arndt, J. (2012). The Power of the Past: Nostalgia as a Meaning-Making Resource. *Memory, 20*(5), 452–460.

Rushdie, S. (1991). *Imaginary Homelands—Essays and Criticism 1981–1991*. Granta in association with Penguin.

Said, E. W. (1984). *Orientalism* (2nd ed.). Vintage Press.

Scannell, L., & Gifford, R. (2010). *Journal of Environmental Psychology, 30*(1), 1–10. https://doi.org/10.1016/j.jenvp.2009.09.006

Sedikides, C., & Wildschut, T. (2018). Finding Meaning in Nostalgia. *Review of General Psychology, 22*(1), 48–61. https://doi.org/10.1037/gpr0000109

Sinn, E. (1989). *Power and Charity: The Early History of the Tung Wah Hospital, Hong Kong*. Oxford University Press.

Somerville, P. (1992). Homelessness and the Meaning of Home: Rooflessness or Rootlessness? *International Journal of Urban and Regional Research, 16*(4), 529–539. https://doi.org/10.1111/j.1468-2427.1992.tb00194.x

Tanselle, G. T. (1998). *Literature and Artifacts*. Bibliographical Society of the University of Virginia.

Tuan, Y.-F. (1974). *Topophilia: A Study of Environmental Perception, Attitudes and Values*. Prentice Hall.

Watson, R. S. (1988). Remembering the Dead: Graves and Politics in Southeastern China. In J. Watson & E. Rawski (Eds.), *Death Ritual in Late Imperial and Modern China* (pp. 203–227). University of California Press.

Wildschut, T., Sedikides, C., Arndt, J., & Routledge, C. (2006). Nostalgia: Content, Triggers, Functions. *Journal of Personality and Social Psychology, 91*(5), 975–993.

CHAPTER 7

The Concept of Co-creation and Yim Tin Tsai

Abstract The concept of co-creation originated in business management as a form of economic strategy that brings different parties together (for instance, a company and a group of customers), in order to produce jointly a mutually valued outcome (Prahalad and Ramaswamy in *Journal of Interactive Marketing* 18:5–14, 2004). It has since been applied to a wide range of social, cultural and environmental situations where, in the context of heritage-making and tourism, creative outcomes have been achieved through cooperative interaction between communities, artists (as both individuals and organizations), commercial entities and visitors. Co-creation has been a key factor in the pilgrimage-focused rejuvenation of Yim Tin Tsai.

Keywords Co-creative partnerships · Collaboration · Hulu culture · Hong Kong Tourism Commission Arts Festival · Haptic technology · Virtual space · Symbiotic mutualism

The concept of co-creation originated in business management as a form of economic strategy that brings different parties together (for instance, a company and a group of customers), in order to produce jointly a mutually valued outcome (Prahalad & Ramaswamy, 2004). It has since been applied to a wide range of social, cultural and environmental situations where, in the context of heritage-making and tourism, creative outcomes have been achieved through cooperative interaction between communities, artists (as both individuals and organizations), commercial entities and visitors.

With its economy originally based on a 2000-year-old traditional way of salt-making using seawater, and the community's conversion to Catholicism in the 1860s, the combined cultural/industrial heritage of Yim Tin Tsai is unique for Hong Kong. Its distinctive history is bringing the island back to life and it owes its revitalization to manifestations of co-creation among key stakeholders. As applied to heri-ligious tourism in this instance, the original Hakka village community combined with the wider Catholic community of Hong Kong to initiate pilgrimage, a sacralized heritage conservation project that has resulted in the renaissance of their village as a satisfying religious and cultural experience for pilgrims and other visitors to the island. Their cooperative effort is a textbook 'fit' with the touristic definition of co-creation as *"a shift from top-down solution-building to joint and collaborative processes whereby people and organisations generate solutions, build capacity, and create value together"* (Phi & Dredge, 2019).

An eight-point structure of co-creative tourism (after Sofield et al., 2017; Richards & Marques, 2012) identifies the main elements of the cooperative basis for new pilgrimage tourism in Yim Tin Tsai's renaissance:

    i. A source for re-creating & reviving sense of place;

    ii. A form of clan expression, community identity, and solidarity;

    iii. A symbolic intervention by an international 'outside' agency (the Vatican) that generated the inclusion of the wider Hong Kong Catholic community into the affairs of Yim Tin Tsai and led to the alliance with the Hakka Chan clan;

    iv. A creative means of using existing resources;

    v. A means of strengthening place identity and distinctiveness for Hong Kong, especially through the canonization of Father Freinademetz and the two heritage awards granted by UNESCO to the Yim Tin Tsai conservation projects;

    vi. A form of self-expression/re-discovery initially for the Yim Tin Tsai diaspora, then the Catholic community of Hong Kong, now extended to other residents of Hong Kong who through visitation uncover a hidden part of their history;

vii. A form of 'edu-tainment', self-realization and education; and
viii. The creation of 'atmosphere' for a place that is starkly different
from the towering skyscraper cityscapes of Hong Kong, most
recently exploited by Tourism Hong Kong on Yim Tin Tsai in a
specific instigation of co-creation in 2019–2022.

The emergence of co-creative partnerships in Yim Tin Tsai is founded
on community efforts at place-making, focusing initially on built
heritage—the restoration of its ruined church, the transformation of the
derelict primary school into a folk museum and Exhibition Centre, and
the reconstruction of the abandoned salt pans in which many volunteers participated. Since then, with the advent of pilgrimage and its
associated ceremonies, the provision of traditional Hakka meals by volunteers for group tours, the participation of visitors in harvesting salt (at
different stages in the evaporation process, visitors are invited to open
and close seawater 'gates' leading to the ponds, rake the salt crystals over
and transfer the final product into vats), the Catholic Nature Trail of
Reconciliation and the interpretation of the island's former Hakka village
society by volunteer guides, it has incorporated intangible heritage into
its repertoire of attractions.

Where most CBT welcomes visitors to observe in real time the normal
everyday life of its village inhabitants going about their daily activities,
Yim Tin Tsai does not—although the guided tour of the Cultural Exhibition Centre with its accompanying narration and displays of artefacts
and villagers' belongings provides an informative introduction to Hakka
village culture. However, the constant comings and goings of volunteers (a number of whom are former residents), engaged in various
village-based activities as service providers with participant visitors in tow,
removes it from being a relatively static spectacle to what Tourism Hong
Kong has termed: "A living museum".

The Joint Organizing Committee has been active in pursuing additional co-creative experiences. In 2016, for example, it invited the Hong
Kong non-profit organization, Hulu Culture, to add artistic elements
to enhance the heritage experience of a visit to the island. One result
was a series of mosaic artworks installed in the windows of various
derelict buildings, depicting cultural scenes from Hakka traditions and
also Catholic ceremonies (see Plate 7.1). Participant stakeholders included
Hulu Culture, a group of Hong Kong artists, the Village Committee,
the Catholic Diocese of HK, the HK Jockey Club Trust established to

sponsor public art, and former residents who gave approval for their houses to be used for the project. Hulu Culture was a particularly appropriate agency to undertake the venture because of its involvement in preserving Hong Kong's traditional culture since its inception in 2004. Its name is derived from a Chinese myth in which one of the Eight Immortals of Daoism, Li Tieguai (Iron-Crutch Li), had a calabash or *hulu* from which he dispensed magical cures for poor and oppressed people. The NGO has been energetic in helping local authors and historians publish their works, and has held a number of exhibitions with community culture as their themes, in which artists, scholars and students have been invited to participate and share their expertise (http://www.huluhk.org/pdf/HuluLeaflet).

The sculptures created in the first three Hong Kong Tourism Commission-sponsored arts festivals were an admix of Christian, Hakka and Nature themes, a number of them with religious meaning incorporated into the Trail of Reconciliation as more or less permanent fixtures. They include the *"Wall of Sanctification"*, a stone bench and perforated steel wall created as a memorial to honour the pioneer evangelists and located in the ruins of Father Freinademetz's original chapel (see Plate 7.2); and a mural created by a nun on the wall of another abandoned house adjacent to the cottage occupied by Father Freinademetz called *"Everything is connected: Laudato Si Wall"*[1]—taken from Pope Francis' 2015 Encyclical on the environment—'*On Care for Our Common Home*'.

Because of COVID-19 restriction on movements, details of the 2019–2021 festival artworks were posted via a 360-degree virtual reality function with audio guides on Tourism Hong Kong's official website. In its promotional material, Tourism Hong Kong described the arts festivals as *"an active exercise in co-creation …. that turned Yim Tin Tsai into an "open museum", bringing a new and unique travel experience integrating arts, religion, culture, heritage and green elements to visitors"* (https://www.info.gov.hk/gia/general/201911/30/P2019112900769.htm, accessed 26 September 2021).

With the easing of COVID-19-related restrictions on internal travel within Hong Kong in 2022, a second three-year set of Arts Festivals was initiated by the HK Tourism Commission and expanded to cover

---

[1] *Laudato Si* means *"Praise be to you"* which is the first line of a canticle that praises God for all of his creation, written by the thirteenth-century friar, St. Francis, who founded the Franciscan Order and is the patron saint of ecologists.

Plate 7.1   Co-creation in Yim Tin Tsai (A series of ceramic mosaics were created in 2016 and installed in the empty window frames of derelict houses, reminiscent of stained glass windows. They capture both religious events and Hakka cultural events, portraying the heritage of everyday living in the village before it was abandoned)

artworks on four additional islands near Saikung. Yim Tin Tsai retained its central role, and again the Joint Organizing Committee was able to reach agreement on showcasing its religious, Hakka traditions and natural heritage: artists were invited to work with themes like village stories, rural and urban balance, ecological conservation and cultural inheritance. The new festival began with a week of interactive tours and events on Yim Tin Tsai in November 2022 with six new artworks, two of which expressly relate to Christianity. One of them, *Homeward Voyage*, is located next to

Plate 7.2   Christian themed artworks—expressions of symbiotic mutualism

the pier below the church which stood out on the hill like a beacon for returning fishermen. Composed of two doves representing a boat and sail, this sculpture signifies the Holy Spirit's guidance and protection on every boat's journey home (see Plate 7.2). A second sculpture, located near the mid-point of the Nature Trail of Reconciliation, is titled "*Everything is in God's hands*" and represents the outline of two hands laid out as a maze for visitors to reflect on the joys and sorrows of life as they journey to the inner core of the maze. The site of this sculpture is on the former sports ground for the village school and adjacent to the Catholic cemetery, signifying the joys of childhood and the sorrows of death, thus capturing the sacred/secular gaze simultaneously (see Plate 7.2). For former residents of Yim Tin Tsai for whom this site was a childhood playground, its semiotic aura is magnified through their collective memory and nostalgia for 'a golden time long past', beyond the current spiritual messages embodied in the physicality of the artwork.

Two other artworks in particular deal with entanglement in a symbiotic trilateral relationship that combines Hakka culture with its sense of place and home, Catholicism, and Yim Tin Tsia's salt-making tradition. The first, entitled "Salt Farm, Catholicism and Hakka" by Wong Chi-chuen, is a kinetic installation *"visually connected via a grid pattern through which the audience can visualise a salt farm, a crucifix and the window of a house. The work intends to connect the wisdom, culture and belief of Yim Tin Tsai's Hakka village in a unique experience that briefly introduces the history of the community. The display shelf attempts to exhibit artifacts in a different way through haptic technology and display screens"* (Hong Kong Tourism Commission web page on its Arts Festival—www.yimtintsaiartsf estival.hk/interest.php?id=3&lang=en). It is installed in a small renovated house below the former primary school-now-folk museum (see Plate 7.3). This display moves beyond Bhaba's 'interstitial space' (Bhabha, 1994), Soja's 'Third Space' (Soja, 1996) and Lefebvre's (2004) 'spatial triad', i.e. perceived space, conceived space and lived space, to encompass 'virtual space' (Kosari & Amoori, 2018, p. 163). It is also a demonstration of co-creation since the observer who uses the haptic technology[2] of this display generates through his/her personal movement (kinetic corporeal energy) the images perceived by him/her. It is a cognitive experience as much as it is a physical encounter with the complexity of the interpenetration (entanglement) of the three themes on which the artwork is based, where harmony rather than dissonance characterizes this interpretation and representation of Yim Tin Tsai (see Plate 7.3). In its own way since virtual space has no physicality this exhibit may be seen as an attempt to embrace the spirituality of the salt fields that is the essence of the islanders' relationship to their heritage.

The second artwork, titled "The Sublime Region" by Ho Chun-kit, Ho Yik-chee, Lee Kong-fai, Leung Ching, Ng Ka-ming, Pong Yuen-kiu and Wong Yat-yau, is also designed to function as a representation of the community's identity, sense of place and spirituality. It is a smaller replica of the brass bell of St Joseph's Chapel that the artists installed hanging from a brass arch behind the church, and captures both the acoustics and

---

[2] Haptic technology transmits tactile information using sensations such as vibration, touch, and force feedback. Virtual reality systems and real-worth technologies use haptics to enhance interactions with humans. One of the goals of haptics is to allow a virtual reality system to make humans feel as if the experiences it portrays are 'real' (Ashtari, 2022).

Plate 7.3   Symbiotic Artworks Embracing Hakka identity, culture and religion

visual design of the original. In their words: "*The original bell used to anchor itself at the heart of Yim Tin Tsai, almost demarcating a sanctified zone within the sound of the bell as it summoned villagers to mass at the chapel. If you walk under the installation, you can hear the bell, feel the Lord's charity, and raise your spirit and faith. The piece pivots the replica bell as a nexus for the trinity that is God, Yim Tin Tsai and villagers, and links up Sky, Earth, and Humans in a parallel analogy*" (Hong Kong Tourism Commission web site: www.yimtintsaiartsfestival. hk/interest.php?id=3&lang=en, accessed 4 April 2023). This artwork also emphasizes and promulgates harmony between faith and heritage, the bell and arch a sacralized object that in its placement outside the church is not reserved for followers of Catholicism but is accessible to all (see Plate 7.3).

The sculptures, murals and other works generated by the HK Tourism Commission's Arts Festival that adorn the pilgrims' Trail and other places on the island may be seen as a contemporary expression of the way in which art and aesthetics—material culture, built heritage and artefacts— enhance the richness of many ancient pilgrim trails and religious sites (the *camino* and the cathedral of Santiago de Compostella are prime examples). Di Giovine and Garcia-Fuentes (2016, p. 6) suggest that such "*elements of material culture … are part and parcel of these sites … (and that) aesthetics and religious experiences are inescapably entangled experiences that reinforce and complement one another*". The new artworks on Yim Tin Tsai thus become objects of the 'sacred/secular gaze', inter-secting the sacred and profane at a point where differing interpretations co-mingle. On the one hand, they enhance the religious experience for the pilgrim and on the other the aesthetic experience for the secular tourist, both types of visitors being infused with (divergent) spiritual appreciation. The varied responses by visitors to the Christian-themed artworks reveal a symbiotic relationship—co-existence, mutualism and commensalism—as identified in our concept of heri-ligion.

The artworks have been supported with various interactive events in additional examples of co-creation, such as audience participation in a *"Common prayer with chants from Taizé Community"* led by Father Dominic Chan with a team of Christian youths. He had spent part of his training at the Taizé village monastery in Burgundy, France, whose mission is to promote the unity of Christians, humans and nature; and these themes have been central to his stewardship of the Catholic Church in Hong Kong ever since. This event was designed in his words: *"To open hearts and minds to a sense of reconciliation and communion with ourselves, others and nature through chanting, listening to God's Word, meditating on the beauty of nature and enjoying the artworks surrounded by the healing atmosphere of Yim Tin Tsai"*. The prayers were taken from the Bible and visitors joined with the choir in chanting the verses to immerse themselves spiritually in their natural surrounds.

The Festivals' religiously-oriented artistic endeavours on Yim Tin Tsai may be seen as exercises in symbiotic mutualism whereby both key symbionts, i.e. the island's Joint Organizing Committee and the HKTC, were able to reach an outcome that served both of their purposes without engendering friction. The combined endeavours of the Catholic Church and the Village Committee to establish "Salt and Light" for the manage-ment of its restoration projects and its pilgrimage trail, coupled with the Festival's Christian-themed artworks, represent the site's 'spiritual magnetism' (Preston, 1992).

In short, Yim Tin Tsai represents a rich and diverse 'laboratory' of aspects of co-creation, including community activism, co-created social innovation, cross-sectoral co-creation, learning, capacity-building and knowledge dynamics, co-created pilgrimage and tourism policy, gender equity issues, and power and politics, among others. The partnership between the Hong Kong Tourism Commission and the Joint Organizing Committee reflects symbiosis, the artistic outcomes of which may be seen as 'heri-ligion' in action.

References

Ashtari, H. (2022, November 4). *What Are Haptics? Meaning, Types, and Importance*. Spiceworks. https://www.spiceworks.com/tech/tech-general/articles/what-are-haptics/

Bhabha, H. K. (1994). *The Location of Culture*. Routledge.

Di Giovine, M. A., & Garcia-Fuentes, J. M. (2016). Sites of Pilgrimage, Sites of Heritage: An Exploratory Introduction. *International Journal of Tourism Anthropology*, 5(1–2), 1–23.

Kosari, M., & Amoori, A. (2018). Thirdspace: The Trialectics of the Real, Virtual and Blended Spaces. *Journal of Cyberspace Studies*, 2(2), 163–185.

Lefebvre, H. (2004). *Rhythmanalysis: Space*. Continuum.

Phi, G., & Dredge, D. (2019). Critical issues in tourism co-creation. *Tourism Recreation Research*, 44(3), 281–283. https://doi.org/10.1080/02508281.2019.1640492

Prahalad, C. K., & Ramaswamy, V. (2004). Co-creation Experiences: The Next Practice in Value Creation. *Journal of Interactive Marketing*, 18(3), 5–14. https://doi.org/10.1002/dir.20015

Preston, J. (1992). Spiritual Magnetism: An Organizing Principle for the Study of Pilgrimage. In A. Morinis (Ed.), *Sacred Journeys: The Anthropology of Pilgrimage* (pp. 31–46). Bloomsbury Academic Publishing.

Richards, G., & Marques, L. (2012). Exploring Creative Tourism: Editors Introduction. *Journal of Tourism Consumption and Practice*, 4(2), 1–11.

Sofield, T. H. B., Guia, J., & Specht, J. (2017, August). Organic 'Folkloric' Community Driven Place-Making and Tourism. *Journal of Tourism Management*, 1–22.

Soja, E. W. (1996). *Thirdspace: Journeys to Los Angeles and Other Real-and-Imagined Places*. Blackwell.

CHAPTER 8

Stakeholder Analysis

Abstract Stakeholder theory in its original form focused on the responsibility of organizational managements to take ethical and social considerations into account when its activities impact multiple constituencies and now encompasses environmental issues, sustainability and community concerns amongst others. It provides a useful analytical tool to categorize all major actors in a given situation and identify the roles they perform in attaining a consensual outcome.

Keywords Primary stakeholders · Secondary stakeholders · Active stakeholders · Passive stakeholders · Types of stakeholders

Stakeholder theory in its original form focused on the responsibility of organizational managements to take ethical and social considerations into account when its activities impact multiple constituencies and now encompasses environmental issues, sustainability and community concerns among others. It provides a useful analytical tool to categorize all major actors in a given situation and identify the roles they perform in attaining a consensual outcome.

When there are simultaneous processes of heritagization of sacred sites and elements and sacralization of 'secular' sites and objects in the opposite direction, "*different valuations, interests, expectations, and sensibilities,*

T. Sofield et al., *Heritage-Making in Hong Kong Through Culture and Religion*, https://doi.org/10.1007/978-981-97-4339-1_8

between different authoritative discourses and hierarchies of value, involving different groups of people with a stake in the specific religious heritage site, object, or practice may be apparent" (van den Hemel et al., 2022, p. 10). In this context, stakeholder theory offers *"a holistic approach or procedure for gaining an understanding of a system, and assessing the impact of changes to that system, by means of identifying the main actors or stakeholders and assessing their respective interests in the system"* (Grimble & Wellard, 1997, p. 175). A widely accepted definition of a stakeholder is: *"Any group or individual who can affect or is affected by a development and/or the achievement of an organization's objectives"* (Freeman, 1984, p. 46). Mitchell et al. (1997) identify *narrow* definitions of stakeholder that focus on the direct relevance of that agency or person to the core interests of a project or agency, whereas *broad* views are based on the premise that a development or a business can be affected by, or can affect, almost anyone. Grimble and Wellard (1997) underline the usefulness of stakeholder analysis in understanding complexity and compatibility problems between objectives and stakeholders. Stakeholder analysis is also derived from participatory methods that seek to integrate the interests and perspectives of all parties regardless of power differentials (Chambers, 1997). While initially applied to business operations and management, it has wider application to socio-political and cultural networks (Parmar et al., 2010), and this is the framework which we apply to Yim Tin Tsai.

Hardy (2001) identified levels of importance of stakeholders and different types of stakeholders, as follows:

i. *Levels of importance (in descending order):*

   a. *Primary stakeholders*
   b. *Secondary stakeholders*
   c. *Active versus passive stakeholders*

(a) Primary stakeholders: are ones who, without continuing participation, the business or project or development cannot survive. By definition, they will invariably be active stakeholders.

The Yim Tin Tsai Village Committee, the Catholic Church of Hong Kong and the Joint Organizing Committee established by those two agencies are of the first importance in the island's religious and heritage revitalization. The NGO, "Salt and Light", also falls into this category of stakeholder with its daily management of the island as a destination. They are all characterized by positioning and sponsoring the Catholic heritage of the community as the primary force behind their actions and activities, hence sacralization. Through its support and funding of the two three-year Arts Festivals, the Hong Kong Tourism Commission also assumed the role of a primary stakeholder, but its vision for Yim Tin Tsai as a general tourism destination moves away from the transcendental, religious values espoused by the four stakeholders identified above to focus on 'secular' this-worldly cultural values, even when its support has resulted in religiously-oriented works of art as outlined in the previous section on co-creation. The provision of essential services and infrastructure (the wharf, electricity, water) by the HK Government emphasizes its role in this regard. Since the venture relies upon visitation, pilgrims and general tourists may similarly be considered as primary stakeholders, and those villagers who operate ferry services to provide essential access are also indispensable.

(b) *Secondary stakeholders* are ones who influence, affect or are influenced or affected by the activity but who are not necessarily engaged in direct transactions with the business or project or development and are not essential for its survival. Many of them will be relatively **passive** stakeholders.

The Vatican in Rome, UNESCO, and the Hong Kong Government, through its overarching policy framework related to 'indigenous villages' and other facets of Government and administration, may be considered in this category of stakeholders. Strong differentiation of their roles is apparent. The Vatican, motivated by its spiritual mandate, power and authority to care for Catholics globally, canonized Father Freinademetz with almost certainly no comprehension of the direct and singular impact its pronouncement would have on Yim Tin Tsai. In addition to unwittingly providing the key motivation for the restoration of the Church and the establishment of pilgrimage tourism, the Vatican continues to affect what happens on Yim Tin Tsai by its global coordination of special days of worship, such as All Saints Day.

UNESCO's assessment of the heritage value of the restoration of the chapel provides a degree of ongoing global recognition of Yim Tin Tsai, but as noted above its technical appraisals are based on historical, architectural, aesthetic and cultural factors devoid of a capacity to evaluate what might be called the spirituality of a site. UNESCO's grant of an Industrial Heritage Award for the restoration of the Salt Pans provides further evidence of that organization's role in the 'secular' cultural development of Yim Tin Tsai (noting that the very act of such an award places the site or object into a category of 'outside the ordinary' to become an object of awe in what has also been referred to as sacred without connotations of any specific religion, invoking new forms of meaning making and sacralization: Turner, 2006; Salemink, 2009). At the actual time of its Heritage assessments UNESCO could be acknowledged as a primary, active stakeholder, but once those interventions ended in granting awards it has become a secondary passive stakeholder, maintaining just a watching brief over the restoration of the sites. Hulu Culture mirrors the changing roles of UNESCO in that its initial involvement in co-creation categorized it as a primary active stakeholder, but it is now a passive secondary stakeholder.

The Hong Kong Government's interests in Yim Tin Tsai are primarily regulatory and administrative, enacted under legal provisions such as the 2016 Act on Small House Policy and Planning for Indigenous Villages (revised from the original Act of 1972) and the Rural Representative Election Ordinance 2014, which amended previous Acts to elect village committees, and through its agency, the Hong Kong Tourism Commission, it has played a more direct role as a primary stakeholder as well.

Pilgrims and tourists who both affect and are affected by other stakeholders can be defined as fitting into this second category of stakeholders as well as the first category. Volunteers, on the other hand, whose numbers and activities on any given day depend upon the type and numbers of visitors, may be defined as primary stakeholders.

(c) Active versus passive stakeholders: Many of the villagers who are now Hong Kong residents will be active to varying degrees in events on Yim Tin Tsai, whereas most of the Yim Tin Tsai diaspora may be considered as generally passive: the latter retain an interest in the island as absentee landholders to a greater or lesser degree but their overseas residence and infrequent return for the most part reduce their capacity to interact directly. Some have contributed financially to the restoration of the church, but the pervasive inertia of many nevertheless directly affect some of the physical and material assets of the island as their former residences fall into ruin and are not restored or made available for restoration by others. Diagram 8.1 illustrates the different levels of stakeholder participation in the heritage-making development of Yim Tin Tsai.

(d) Types of stakeholders: Stakeholder literature commonly categorizes types of stakeholders according to their various functions, as follows:

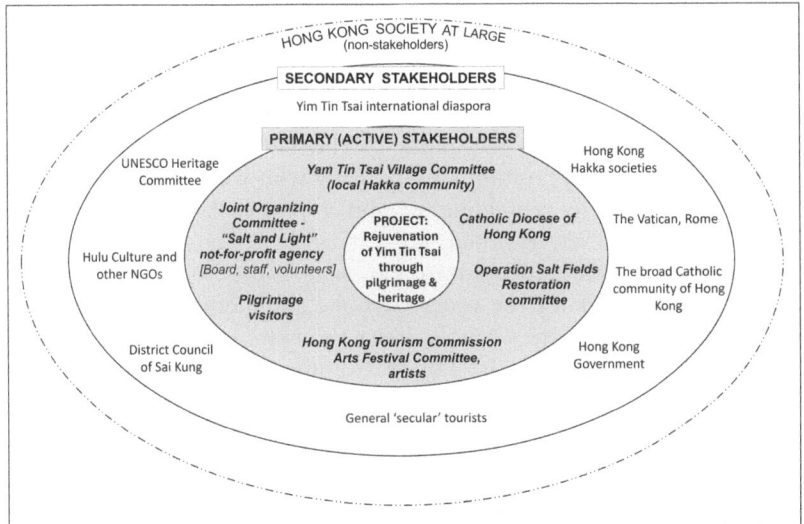

Diagram 8.1 Stakeholder Analysis, Yim Tin Tsai

- Advocates—idea creators, designers
- Champions—to lead the change
- Supporters—"critical mass"
- Decision makers (to approve the activation of critical change enablers—policies, funds, people)
- Opinion leaders—decision-swayers, special interest groups that sustain people's awareness of the need for change (lobby groups, NGOs, media)
- Expert enablers—this is a new type we have added, as explained below

With the demise of the village of Yim Tin Tsai in the late 1990s and a moribund Rural Village Committee, a breath of fresh air was introduced into the situation with the election of a new relatively young chief, Colin Chan Chung-yin, in 1999. His 'platform' was and remains what has become an abiding mission: to resurrect the island's heritage to restore the village. In Hardy's (2001) terms, he was clearly an 'advocate'. For the first few years, he focused on building a network of Chan villagers from Hong Kong and around the world, hoping to create a like-minded community of descendants who would help to rebuild Yim Tin Tsai in a sustainable way. Chan set an example by restoring his ancestral home, although he remains resident in Sai Kung. But he made only slow progress, since without an over-riding reason for former villagers to re-invest in any restoration of Yim Tin Tsai other than general nostalgia, there was opposition among many who were concerned about the costs involved in rebuilding it, and a passive lack of interest by others. Then, however, as noted, the Vatican in Rome became a crucial catalytic agency with its decision to canonize Father Freinademetz, and this was the stimulus for the wider Catholic community of Hong Kong through the agency of the Catholic Vicar General of Hong Kong, Father Dominic Chan, to partner the Village Community in a drive to restore the dilapidated church.

At this stage, Colin Chan transitioned from being a visionary or advocate to 'a local champion' utilizing his position within the Village Committee to forge a formal relationship with the Catholic Church of Hong Kong to enable development to proceed. Father Dominic Chan may also be titled a 'champion' as he was the driving force behind the fund-raising effort to restore the chapel as a permanent memorial to the newly-canonized Saint Freinademetz. The actions of these two key individuals in fact encompassed two other stakeholder designations—their pre-eminent positions in their respective organizations automatically cast them in the additional roles of 'opinion leaders' and 'decision makers'. Both carried considerable mass support behind them, i.e. the Yim Tin Tsai community and the Catholic community of Hong Kong, and the fact that Father Chan not only carries widespread respect for his leadership of the Catholic Church in Hong Kong but is also a Yim Tin Tsai Chan clan member and therefore an 'insider', added weight to his representations. To implement the physical restoration of the chapel, technical input from a specialist heritage architect, Anna Kwong, was necessary and we would thus add another category of stakeholder role to the pantheon, that of 'expert enabler' since her expertise was fundamental and essential to the success of the venture.

The establishment of the Heritage Exhibition Centre saw Colin Chan again assume the roles of advocate, opinion leader and local champion as he initiated this community-undertaken activity and steered it through various vicissitudes to fulfilment. Professor Lam and Professor Cheung were also expert enablers whose invited interventions helped to bring about the successful completion of the Heritage Centre project.

When we examine the restoration of the salt pans, we find that the documentary film maker, Wong Tin-Shing became a central agency of the heritage reconstruction effort. His enthusiasm had him undertaking the role of 'advocate' and he then transitioned into a 'champion', incorporating the restoration work on the salt fields into a 12-minute documentary, wearing the hat of 'opinion leader' (we might modernize this label in terms of the contemporary widespread popularity of social media such as Facebook and the former 'Twitter' and term him 'an influencer'). His efforts, however, required the support of Colin Chan, and that was forthcoming. Colin Chan again assumed multiple roles—as champion, as opinion maker persuading the village community to support this new, non-religious venture (Operation Restoration Salt Fields), and decision-maker to raise the necessary funds, appoint technical enablers to implement the project, and then work with "Salt & Light" to oversee its operation and manage its visitation. A retired engineer, David Ip Chan Lap, was an enthusiastic supporter of the salt fields restoration project, volunteering his own time to investigate the *modus operandi* of solar evaporation salt production in Taiwan, supervising technical aspects of the reconstruction on Yim Tin Tsai, and then taking on the role as volunteer manager of the salt fields from their inception until he retired in 2021: in stakeholder theory, an expert enabler. The power and authority of the range of stakeholders within the community to determine a variety of outcomes underline the concept of empowerment as the capacity to take, make and implement decisions of their own accord.

The HKTC may be termed both advocate and enabler since its role in imitating the 2016 Hulu cultural festival with its proposals for three-year Arts Festivals injected the funding necessary for their actualization. Its willingness to embrace the religious/cultural/heritage framework of the Joint Organizing Committee's vision for Yim Tin Tsai has enhanced the processes of heri-ligion that are being played out in this island milieu.

This excursion into stakeholder analysis is not comprehensive, and one avenue of further research to gain a more profound understanding of the dynamics underlying the rejuvenation of Yim Tin Tsai through heri-ligion would be to delve more intensely into the networks and key actors who have been responsible for the current outcomes.

REFERENCES

Chambers, R. (1997). *Whose Reality Counts?* Intermediate Technology Publications.

Freeman, R. E. (1984). *Strategic Management: A Stakeholder Approach.* Pitman.

Grimble, R., & Wellard, K. (1997). Stakeholder Methodologies in Natural Resource Management. A Review of Principles, Contexts, Experiences and Opportunities. *Agricultural Systems Journal. 55*(2), 173–193.

Hardy, A. L. (2001). *A Troubled Paradise: Stakeholder Perceptions of Tourism in the Daintree Region of Far North Queensland, Australia* [Doctoral Dissertation, University of Queensland].

Mitchell, R. K., Agle, B. R., & Wood, D. J. (1997). Toward a Theory of Stakeholder Identification and Salience: Defining the Principle of Who and What Really Counts. *The Academy of Management Review, 22*(4), 853–886. https://doi.org/10.2307/259247

Parmar, B. L., Freeman, R. E., Harrison, J. S., Wicks, A. C., de Colle, S., & Purnell, L. (2010, June). *Stakeholder Theory: The State of the Art.* The Academy of Management Annals.

Salemink, O. (2009). Afterword: Questioning Faiths? Casting Doubts. In T. D. DuBois (Ed.), *Casting Faiths: Imperialism, Technology and the Transformation of Religion in East and Southeast Asia* (pp. 257–263). Palgrave Macmillan.

Turner, B. (2006). Religion and Politics: Nationalism, Globalisation and Empire. *Asian Journal of Social Science, 34*(2), 209–224. https://doi.org/10.1163/156853106777371175

Van den Hemel, E., Salemink, O., & Stengs, I. L. (2022). Introduction: management of religion, sacralisation of heritage. In E. Van den Hemel, O. Salemink, and I. L. Stengs, (Eds.), *Managing sacralities at religious heritage sites in contemporary Europe.* London: Berghahn Books, 1–21. Google Scholar.

CHAPTER 9

Entanglement

Abstract Entanglement describes the interweaving of human–human and/or human-inhuman entities inter-acting at different scales that create complex relationships, with the literature generally emphasizing negative manifestations. Entanglement is always an emergent process whereby relationships develop, dissipate and reconvene, and this chapter identifies some of the examples of entanglement that we have found on Yim Tin Tsai where harmony rather than dissonance has been relatively common occurrence.

Keywords Assemblage · Authorized heritage discourse · Commodification · Durability · Harmony · Dissonance · Trans-liminal public space · Semiotics · Syncretism · Symbiosis · Mutualism

Entanglement describes the interweaving of human–human and/or human-inhuman entities inter-acting at different scales that create complex relationships, with the literature generally emphasizing negative manifestations. Entanglement is always an emergent process whereby relationships develop, dissipate and reconvene, and this chapter identifies some of the examples of entanglement that we have found on Yim Tin Tsai where harmony rather than dissonance has been relatively common occurrence. While entanglement is often used interchangeably

T. Sofield et al., *Heritage-Making in Hong Kong Through Culture and Religion*, https://doi.org/10.1007/978-981-97-4339-1_9

97

with assemblage, current practice locates them in a hierarchical relationship so that an assemblage is the overarching aegis that consists of a
multiplicity of different entanglements between entities. According to
Lisle (2021, p. 1):

> An assemblage is a coming together – an assembling – of different entities
> that stick together – or 'entangle' – in a more or less durable formation
> for a period of time. These entities are not all human or even man-
> made; indeed, assemblages are also constituted by non-human entities like
> objects and technologies, as well as intangible substances like ideas and
> atmospheres. Assemblage thinking is important because it foregrounds the
> relations between the entities that entangle together. It helps us examine
> durability (i.e. how disparate actors co-ordinate, hold together and fall
> apart) and transformation (i.e. how relationships change the entities them-
> selves, and how entanglements mutate over time). … Assemblage is the
> name of the congregation that is made up of multiple entanglements and
> that endures over time.

In its fundamentals then, an assemblage emphasizes and reinforces the
fact that entanglement is always dynamic, a process, because entanglements are always emerging, dissipating and reconvening. Dissonance and
controversy constitute a useful compass to point to tensions that can
arise within and between various human/non-human entanglements and
in this context Actor-Network-Theory (ANT) scholars have developed a
'*cartography of controversy*' as a methodology that uses digital technologies to map the multiple human and non-human actants that entangle
together as they negotiate a '*matter of concern*' (Venturini, 2010, p. 260).
Interrogating dissonance and controversy—or their opposite, harmony—
as entanglements unfold in-process, is particularly useful because such
analysis inevitably reveals a multitude of stakeholders whose motivations
and involvement will be made transparent, and that their relationships
can dis-entangle and re-entangle. In pursuing the complexities of entanglement in the situation of Yim Tin Tai's rejuvenation, we focus on two
defining characteristics: (1) that entanglement invariably involves both
human and non-human entities, and (2) that entanglement is always
emergent and in process.

Because the 'heri-ligion-ization' of Yim Tin Tsai remains strongly
within the control of the Joint Village/HK Catholic Committee where
there is only one clan in the village, all members of the Committee
share the same religious background, and "Salt & Light" charged with

management is entirely within their purview, the island has been able to resist to a significant extent the disruption of pilgrims' devotional behaviour by common secular heritage tourism practices where guides with loudspeakers inveigle their hordes with non-biblical interpretation and 'invade' sacred spaces. In its early years Yim Tim Tsai was basically devoid of the tensions and dissent that have accompanied many cultural heritage village-based projects elsewhere. This is not to say that there has been no internal disagreement within the village community concerning the future trajectory for Yim Tin Tsai. As noted above, the lack of economic opportunities has been a major deterrent for some villagers to invest in resurrecting the village although they tend to support conservation measures that have not created personal financial imposts (Chan et al., 2014). Passivity rather than opposition has characterized their locus. From around 2005 until the advent of the HK Tourism Commission with its Cultural Festival lauding the island as a 'new' tourist attraction after 2019, Yim Tin Tsai enjoyed a single 'authorized heritage discourse' (Smith, 2006) that was based on its Catholic legacy, combined to some extent with indigenous Hakka vernacular meanings and values. The restoration of the salt fields, while obviously not inherently religious, nevertheless fitted into this narrative. Their association with the dominant religious theme was embodied in the adoption of the biblical phrases that Jesus used to personify his followers as 'Salt' and 'Light' through the title of the Island's management body (Matthew's gospel, Ch. 5, recording the details of the 'Sermon of the Mount' preached by Jesus on a hill above the Sea of Galilee). While there are no physical barriers to simply 'dropping in' on the island (which is popular particularly among kayakers and 'boaties' who on weekends number one or more thousand sailing around the Sai Kung islands), non-pilgrimage visitation was originally not encouraged. The restored church was kept locked, and all bookings were channelled through "Salt & Light", which would organize transport to the island through small village-owned ferries (*kaito*), with volunteer guides and staff to open the church, escort groups around the Nature Trail and the salt fields, and open the restaurant when requested by groups for all-day tours. Infrequently, small commercial tour ferries with about 10–20 sightseers would arrive unannounced, and "Salt & Light" would not open the church or undertake guided tours, so itinerant tourists occasionally expressed irritation. "Salt & Light" staff would state that they were free to walk around the deserted village and nature trail and politely refer them back to the tour operator if they claimed that they had been promised on-island services. The nature of these encounters was not common, was low-key, and it would be inappropriate to classify them as anything other than perhaps portending entanglement as non-pilgrimage tourism grew.

With more 'secular' visitors than pilgrims now frequenting Yim Tin Tsai, however, entanglement in its classic application has gained more analytical relevance as social boundaries emplaced by "Salt & Light" give rise to complaints about restrictions on movement. For example, the use of St Joseph's Chapel for religious purposes results in the 'lock-out' of secular tourists when Catholic ceremonies take place (such as Sunday mass, the celebration of St Joseph's Feast Day and other Catholic rituals and services), and a dress code is enforced that bans visitors clothed in swimming and water-sports attire from entering the church at all times. This is an affirmation in practice that events and their spaces become 'sacred' when they are identified with the places where ritual is enacted. As Knott (2015, p. 102) said: *The role of ritual (as an act of sacralization) is crucial in creating meaningful places, as well as in marking out a sphere of difference and thus producing an ecosystem of religious topography*. Tsivolas (2019, p. 288) has termed this *"the right of veto intended, primarily, to protect the sacred dimension of such edifices"*, and this type of exclusion may lead to tension where visitor expectations cannot be met. Originally no real effort was expended in restoring the primary school nor the salt pans, even though the historical importance of the latter was the determinant for the original choice in siting the village 300 years ago and secular tourists freely wandered around them. Now, however, visitation to these two sites requires payment of an entrance fee and this has not been welcomed by some tourists. How these histories and these sites are viewed, with differing expectations raised, leads to dissonance that is typical of entanglement. However, over time, it may be expected that visitors will accept entrance fees as 'normal', friction will lessen, and entanglement will be seen as a process rather than something fixed in time and durable.

By contrast to Yim Tin Tsai's authorized heritage discourse, in China contending voices among stakeholders often include the state, local Government, outside developer interests and the local community (this latter invariably in a relatively subordinate position), and the result is differing heritage discourses coexisting in the one locality (Sofield & Li, 2020; Zhang & Wu, 2016) where vernacular and indigenous values are often ignored. That situation is compounded when there are different clans within the same village often inhibiting a unified community approach to challenge outside interests. The dominant clan may try to utilize its ancestral hall as a focal point for its version of village conservation combined with senior positions in village society to 'steer' the heritagization development in its favoured direction, to the exclusion of other clans and/or minorities. In China, the state has an official, longstanding policy to utilize the 'New Socialist Countryside' policy, Central Committee of the Communist Party 2005. *Inter alia*, it actively encourages investment from outside agencies and in this context a universal approach to heritage sites all over China is to award a developer the monopoly right to charge an entrance fee—the so-called ticket economy—and income so generated is not always equitably distributed to local residents (Sofield & Li, 2020). The co-option of heritage by the State for national tourism development as part of its efforts to frame a narrative on nation-building often leads to social and economic tensions between big business, local communities and government officials where the power imbalance invariably operates against grass-roots interests. Issues arising from such entanglement induce contestation and dissonance in China.

The relative harmony and lack of tension during the first decade or more of Yim Tin Tsai's heritage development has the potential to erode, however, following the HKTC's arrival on the scene with its grants to establish the Arts Festivals. Its remit *inter alia* is to foster heritage, culture and the natural attractions of Hong Kong and it has been prominent for many years in promoting a diversity of cultural events such as the annual traditional dragon boat races (Sofield & Sivan, 1994), the annual Cheung Chau Island Daoist *Da Jui* ('bun festival'), and the Chinese New Year dragon dance festival not simply in terms of their cultural authenticity and integrity but to promote financial returns from tourism. While it pursues a cooperative relationship with organizations responsible for such events where they retain oversight of their activities, the HKTC is not necessarily averse to a degree of 'Disneyfication' and inclusion of ancillary services and operations that may detract from the actual festival or event. It may also seek to present its own narrative that varies from the narrative pursued by the responsible management regime, and issues of entanglement begin to emerge. This is consistent with di Giovine and Garcia-Fuentes (2016, p. 10) who state that heratigization "*changes a religious relic into an artefact*".

Such an approach conforms with Ashworth's view, who states that for interpretation of heritage, there are "*quite different ways of viewing the past from the present*" (2011, p. 1). Father Freinademetz was only with the parishioners of Yim Tin Tsai for less than three years (1879–1881), but after he became famous in the 1930s, his presence was referenced by the villagers as a point of historical interest about their own Catholic history. Nonetheless, this had little import beyond the narrow confines of Yim Tin Tsai since his international fame rested on his subsequent work during two decades in Shandong Province in China. However, with the elevation of Father Freinademetz to sainthood in 2003, that very slight, localized, historical moment was re-interpreted to give new value to his presence in Yim Tim Tsai as of enduring global significance. St Joseph's Chapel, as evidence of those past connections, became the focal point for heritage conservation. Until the papal decision by the Vatican, the run-down church on Yim Tin Tsai was considered of no particular value by the Hong Kong Catholic Church and, as noted above, was slated for de-consecration, its upkeep too expensive to be justified. The contemporary reinterpretation of this particular piece of history and the value to be accorded to it, however, placed the restoration of the church above all other considerations. Subsequently as it became apparent that support structures were needed to allow the restored church to function effectively as both a pilgrimage site and a heritage site, restoration efforts were expanded to include upgrading of the wharf (funded by the Government), a village house to become the HQ for the NGO, "Salt & Light", two other houses as café/restaurant facilities, and the prayer house as noted above.

These developments denote the way in which heritage is often focused on relict-built miscellanies and the way in which different stakeholders may manifest *polarized views of place planning and management,* with their endeavours resulting in adaptive re-use that may fit more comfortably with some visitors than with others (Ashworth, 2011, p. 2). While the church continues to function as a place of worship in the context of pilgrimage, for the annual commemoration of St Joseph's Feast Day in May, and All Saints Day in November, it no longer holds weekly services but it is visited often many times a week as a dual heritage tourist-and-pilgrim site. There is thus some modification to its purpose since the experience of worship is not always present in visitation by general tourists nor is it used just for local village purposes since its new international profile now acts as a beacon to draw both religious and non-religious visitors to its doors. These new activities and visitation have not compromised in a substantive way the authenticity of the church in terms of original built structure and provision of religious services. Its past is intrinsic to its present, linked in an imagined way.

With reference to the separation of sacred and secular space on Yim Tin Tsai through its management regime, until 2020 "Salt & Light" focused more or less exclusively on pilgrimage visitation and the office was only open on weekends except for pre-booked tours. As mentioned, the church and folk museum were locked for non-pilgrimage visitors who 'simply arrived' on the island without bookings. Just prior to the 2021 Arts festival, and COVID-19 restrictions on movement were eased, "Salt & Light" modified its operation into 6-days-a-week (closed only on Mondays) so currently the chapel and folk museum are locked for only one day a week. While entrance to the chapel is free, it will be reserved sometimes for spiritual activities only, since as one volunteer said: "*Religious purpose is the first priority for the village community. I often see some tour groups feeling disappointed because their schedule clashes with the religious event and they are unable to enter the chapel within their period of visit. But this arrangement is deemed necessary and it's a deliberate policy to retain its sacred status*". This comment is in contrast to one finding of the Danish Heri-ligion team when they investigated the Danish World Heritage site of Roskilde cathedral where the resident priest remarked: "*We limit our congregation members' rights in order to accommodate things like guided tours*" (Salemink et al., 2020, p. 84).

In the context of Yim Tin Tsai, once the Church was granted a Heritage award by UNESCO on technical, aesthetic, historical and architectural conservation grounds but not for any special religious reasons it inevitably became a spectacle (Salemink, 2016) for the secular tourist gaze. UNESCO's Heritage award is focused on material heritage, and the spiritual domain that is so fundamental to making Yim Tin Tsai a special place for Catholics and former residents, played little part in its architecturally-oriented assessment of the heritage conservation values of the restored building. This 'secularization' of St Joseph's Chapel by UNESCO, coupled with Tourism Hong Kong's promotion of Yim Tin Tsai as an interesting and attractive place to visit, was further reinforced by UNESCO's industrial heritage award for the restored salt ponds, especially with their promotion to school groups for educational tours that are divorced from religious motivation. What is eventuating is 'co-existence' of religious and secular visitors vying for access to the same spatial elements. Bremmer (2006, pp. 25, 30) has termed this duality of perceptions of space as '*parallel geographies*' and a '*simultaneity of places*' which is co-habited by *'both heritage/tourist and religious/pilgrim'* (cited in Salemink et al., 2020, p. 83). While all visitors are free to wander through the village and walk along the Trail, they must now pay a fee to visit the salt fields and folk museum. The church is open for general visitors six days a week free of charge provided they do not wear water sports/beach bathers' apparel as noted, but on Sundays the church is open only to worshippers with wardens monitoring entry. The office for "Salt & Light" is open for the purchase of souvenirs there, but the church remains free of such commercialized operations as a reflection of the biblical account of Jesus driving the merchants out of the temple in Jerusalem (The Bible, Gospel of St John, ch. 2, v. 15–16; Matthew, ch. 21, v. 12–13), a deliberate policy designed to emphasize its sacred status. This non-monetized philosophy stands in contrast to many other heritagized religious sites where commodification has been accepted and actively introduced. As the most famous example among many, where religious functionaries argue that commercial activity is a necessary intervention into sacred affairs, the Vatican in Rome has a series of tickets ranging from $6 to $175 for different tours. In 2022 the complex was visited by 5,080,866 persons, 215% more than in 2021, but still below pre-COVID-19 attendance and raised more than $62 million (*The Art Newspaper Survey*, March 2023). The €5 entrance fee to the World Heritage Site of Saint Mary's Catholic Cathedral in Lisbon is justified by the church authorities as necessary to assist in the very significant maintenance costs of the complex. Its oldest buildings date back to the twelfth century, and more than 500,000 visitors (as distinct from worshippers who attend its regular services) entered the complex in 2019 before COVID-19, raising more than €2.5 million. Yanata and Sharpley (2021) describe how financial need underpins the temple-stay experience at the Koyasan Buddhist Mountain temple in Japan but which is presented as an outreach opportunity whereby the monks can fulfil their mission to teach the Buddhist credos—*a commodifying religious praxis*, in the words of Thouki (2022, p. 1043). The various case studies of the EU HERA project include a number of similar examples of enterprise culture adopted by European religious institutions. Heritagization and commodification may thus transform religious sites into what Vukonic (2002, p. 7) described as *'a convenient symbiosis'* where tourist-generated revenue subsidizes the mission of a church or temple. In Yim Tin Tsai, however, a demarcation line has been drawn between the sacred and the secular: for pilgrims, "Salt & Light" 'has constructed an exchange process' where *'instead of money being passed between parties, religious teachings and feelings are exchanged'* (Olsen, 2003, p. 101). For non-pilgrims, the entrance fees for the folk museum and for the salt pans constitute a direct commodification of heritage. For as long as this policy with spatial limitations and deference to religious sensibilities is retained, the tensions that can arise when secular tourists and worshippers share the same space and worshippers are subjected to the tourist gaze as a spectacle (as occurs in many cathedrals in Europe, including St Peter's Basilica in the Vatican in Rome, e.g.), then elements of entanglement will be evident. However, one may anticipate that as secular visitors outnumber pilgrims the management of Yim Tin Tsai may have to make adjustments to such a circumstance.

As with all the artworks on Yim Tin Tsai derived from biblical themes, we have the sacred and secular entangled but for the most part harmoniously so, open and accessible equally to believers and non-believers, subjected to both sacred and secular gazes simultaneously, examples of symbiosis and mutualism. These artworks are in public places not designated as sacred but their overtly religious content inevitably entangles the secular and the sacred in what Van der Tol and Gorski (2022, p. 495) term: *"trans-liminal space: spaces which can contain multiple and potentially conflicting ascriptions of meaning"*. They note that describing public space as trans-liminal *"allows for contemporaneous and competing ascriptions of the secular, the sacred, the secular-sacred, the sacred-secular, without being exclusively grounded in either"* (2022, p. 495). With reference to this viewpoint, entanglement that engenders dissonance may be discerned on Yim Tin Tsai in the Nature Trail and the differing signage that offers both sacred and secular interpretations of the same object. For example, the original signage for the restored village well, erected by the Catholic stakeholders, focuses entirely on the biblical quote signifying Jesus as "the water of life". Adjacent new signage installed in 2021 by the Tourism Commission refers to the "Spring of Living Water" and recounts its 300-year history as the original source of water for the village, later supplemented with a small reservoir on nearby Kai Sai Chau and a 1966 Government-constructed piped flow from the mainland—an historical, heritage-oriented (secular) interpretation that is devoid of spiritual connotations. The very design and physical structure of the signs emphasizes, through a semiotic analysis, their entanglement. In semiotics, a sign is anything that communicates a meaning that is not the sign itself to the interpreter of the sign, in Sausurre's seminal theory (1916), the form of the sign as 'signifier' and its meaning as the signified, or Pierce's triadic relationship between something that stands for something, to someone in some capacity (both authors cited in Danesi & Perron, 1999). The original Catholic sign for the well has its text inscribed on a concrete simulation of the pages of a thick, open book supported on an old wooden post. Its form will be recognized by Christians as signifying the Bible resting on the cross, thus leading to the deeper interpretation of Jesus as life everlasting through his resurrection. The Tourism Commission's sign is an abstract angular aluminium shape mounted on a slim blue post, its form signifying post-modernity. For pilgrims as interpreters of these two signs, their knowledge of Christianity will allow them to understand the deeper meaning behind the original sign, and they will make the link between the title of the new sign ('The well of life') and the original biblical sign; it has a spirituality for them. For 'secular' visitors the linkage will be less obvious and they may fail to recognize the biblical symbolism in the shape and form of the original sign. Semiotically, the well has transitioned from being: (i) purely secular (functional) as the village's source of water, to (ii) sacred as part of the Catholic Trail of Reconciliation, and then (iii) contemporaneously redefined by the Tourism Commission as industrial heritage, to (iv) arrive at 'trans-liminal public space' embracing both the sacred and profane. As Eliade (1959, p. 12) wrote in his seminal work on the nature of religion: *By manifesting the sacred, any object becomes something else, yet it continues to remain itself, for it continues to participate in its surrounding cosmic milieu. A sacred stone remains a stone ... but for those to whom a stone reveals itself as sacred, its immediate reality is transmuted into a supernatural reality.* The village well, now restored, is a vivid example of the sacralization of secular heritage (see Plate 9.1).

In the context of entanglement related to heritage an increasingly fraught issue may be discerned to be developing between the desire and expectations of (secular) tourists to experience the 'ghost village' ruins festooned with creepers and the aerial roots of strangler figs (*vide* MacCauley's 1966 "romance of ruins") contrasted with the modernization of some former derelict buildings. Plate 9.2 with its 'before' and 'after' restoration of a village house that has been modernized and renovated for rental accommodation provides a graphic example of the different aesthetics in play. This building was originally the village store (which occupied the ground floor with the family's residential quarters on the second floor). One villager recalled: "*We have fond memories of Uncle Po's Store. Located next to Ching Po School, the store sold soft drinks and*

Plate 9.1 The semiotics of signage: 'sacred' and 'secular' signs for the village well: trans-liminal public space along the Nature Trail

Plate 9.2 Renovation on Yim Tin Tsai: before and after. **a** Uncle Po's store in dilapidated condition. **b** After renovation as a holiday rental. **c** Sacralized eastern wall, showing juxtaposition with adjacent ruin

candies, among other items. They also sold something called "lottery draw", which was basically a big scratch-off lottery card. On each card, there were many squares, and when you scratched off each square, you had a chance to exchange it for different prizes, like a popsicle or a soda". It was abandoned for a long time and Colin Chan rented this house from its owner, renovated it and now serves as a holiday house.

It is of interest that while the renovation of this house moves away from the traditional vernacular style of village architecture, it nevertheless firmly centres itself in the Catholic heritage of Yim tin Tsai. On the eastern side wall of the house, as part of the Tourism Commission's Art Festival, a mural has been painted that depicts doves rising above St Joseph's Chapel with a couplet in Chinese, the verses of which translate as:

> May heavenly blessings be bestowed upon Yim Tin Tsai.
> May the light of goodwill brighten all humankind

Linguistically, there is a subtle reference to Hakka tradition in the couplet. For centuries, Hakka culture has been famous for its 'mountain songs', which usually consist of two to four sentences with seven characters in each sentence, and these two sentences reflect that traditional linguistic structure with seven characters each, although the Christian content of the verses is far removed from the love songs and folk stories of old. Uncle Po's old store in its rejuvenated form captures the 'sacred/secular' gaze explicitly, its front façade secular, its eastern wall sacralized, and its juxtaposition in front of an old overgrown ruin confronting the observer with the complexities of place attachment and motivations underlying a community's drive to define and/or re-discover and re-establish its roots ('home') where architectural modernization is combined with religion and religious traditions that are of foundational importance.

It is also an example of what Alexis Thouki (**2022**, p. 1037) has termed *"the gentrification of heritage"* and this raises other issues of entanglement such as what to do with abandoned built heritage in the context of three antithetical pairings: *"'preserve as found'* (i.e. leave in ruins) *vs aesthetic restoration; living tradition vs preventing conservation; and staging and gentrification vs museumification"*. He queries whether tourists *"would draw a line between gentrification and integrity (contextual continuity)? ... Do visitors perceive ageing, weathering or even damage as part of authenticity? What ideas can be evoked by signs of decay (patina, moss or roots entwined with masonry)"*, as evidenced by many of the ruins of Yim Tin Tsai? Bearing this in mind, we might ask whether the current hegemony of preservationist culture with its purposeful commitment to a fabric-based conservation ideology at times leads to a rejection or dismissal of lived values and practices held by a community confronting heritagization, and questions as to whose aesthetics (a value-judged concept in itself) might hold primacy. Harding (**2019**) canvasses these matters when reflecting on the significance of Government protection of Swedish heritage churches as post-Christian sacred spaces where Swedish national (and secular) identity, as defined by the Government, plays the determining role in how protection and conservation will proceed.

Under Hong Kong's Antiquities and Monuments Ordinance (**1976**) indigenous villages such as Yim Yin Tsai are not covered and traditional houses in the village setting, coupled with their cultural landscape, are not protected under current heritage law (Lung, **2012**). In addition, in 1972 the Government enacted the so-called Small House Policy (SMH) which gave every indigenous male born in the New Territories the right to build a house with no restrictions as to whether it could be built by demolishing an existing dwelling with a totally different architectural style. As Lung (2012, p. 138) notes, this policy, guaranteed under the Basic Law legislation introduced in 1997 on Hong Kong's return to China, has put traditional vernacular architecture *"in peril"*: he has argued that *"Vernacular settlements are valuable living heritage that should be on an equal weighting with heritage-listed monuments"*. In effect the lack of appropriate heritage protection legislation means that the buildings of Yim Tin Tsai may be modified by villagers as they wish. Where modernization/restoration has resulted in a building that has lost the patina of age and consistency with its original vernacular architectural features and form, the result can be a loss of the aesthetics of the destination in the eyes of tourists (and others). This perception directly counters the desire of those villagers who wish to see Yim Tin Tsai re-inhabited and returned to life as a vibrant resident community, and who view modernization as a right and a worthwhile objective that enhances the value of their properties: gentrification in a new form is acceptable. Entanglement appears in the symbiotic form of competition where there are differing perceptions of aesthetics and whose perception should prevail.

At a micro-level, this one building in Yim Tin Tsai also reflects tension between UNESCO, which has strict regulations governing the restoration of old buildings, monuments and sites for its heritage imprimatur to be awarded, and its sister UN institution, the UN Habitat Commission. UNESCO's regulations on heritage restoration direct that the architectural form, materials and construction methods must be true to their original assembly to retain authenticity, and failure to follow ICOMOS guidelines will negate their ability to obtain UNESCO heritage-designated approval (UNESCO, 1994, 2013). However, the UN Habitat, whose mission is to promote socially and environmentally sustainable human settlements development and the achievement of adequate shelter for all, has been arguing for about two decades that throughout human history, as knowledge about new and different materials, new and different engineering science and construction techniques emerge, all settlements have been dynamic and change has been a constant. To museumize a particular building or precinct and 'lock' it in a particular period and time with heritage preservation and conservation orders based on contemporary ideas of aesthetics is to deny the constantly changing pattern of human settlement over centuries. At the

UN Conference on Housing and Sustainable Urban Development—Habitat III (October 2016) its manifesto stated that: "*Mindsets, policies, and approaches towards urbanization need to change in order for the growth of cities and urban areas to be turned into opportunities that would leave nobody behind*" (UN Habitat Commission, 2022, np). While acknowledging that there is definitely room for conservation of representative layers of successive human habitation and outstanding sites and that authenticity has a role to play in recording human history, Professor Roger Faye, architect and former Director of UN Habitat, voiced concern that globally thousands of local councils have enacted far too many restrictive regulations in the name of 'heritage preservation'. In the conflict between cultural heritage conservation and urban regeneration, especially social housing for poor and disadvantaged populations, since conservation and restoration are invariably more costly than clearance and reconstruction, balance has been jettisoned (personal correspondence, 2019). Habitat's Human Settlements Programme has been unable to make the progress many would prefer. Knippschild and Zölter (2021, p. 547) agree that cultural heritage can be a burden for urban development. In its own micro way, current sacralization/secularization of the built heritage fabric of Yim Tin Tsai mirrors this tension. Entanglement accurately describes this development.

A recent incident regarding the island was that in May 2023 the Yim Tin Tsai Committee rejected a proposal to be submitted to the UNWTO 'Best Tourism Village Initiative', on the grounds that it did not want to be labelled as a "tourist village" instead of a "Catholic village". The primacy of the sacred over the secular is thus being maintained for the present in a manifestation of what van den Hemel, Salemink and Stengs (2022) have called "*managing sacralities*", that is, "*as a way to interrogate the categories of both heritage and religion*" (p. 9). This is a clear demonstration that to date the local has prevailed over top-down control and decision-making, where the Yim Tin Tsai community has mobilized its aims and exercised its agency despite power differentials between it and other stakeholders.

References

Ashworth, G. J. (2011). Preservation, Conservation and Heritage: Approaches to the Past in the Present Through the Built Environment. *Asian Anthropology, 10*(1), 1–18. https://doi.org/10.1080/1683478X.2011.10552601

Bremer, T. S. (2006). Sacred spaces and tourist places. In D. J. Timothy & D. H. Olsen (Eds.), *Tourism, religion and spiritual journeys*. (pp. 25–35). Routledge.

Chan, E., Lam, T., Shiu, D., & Wong, M. (2014). *Renaissance in Yim Tin Tsai: Catholic Traditions and Salt-Making History define Hakka Village—Varsity*. Chinese University of Hong Kong.

Danesi, M., & Perron, P. (1999). *Analyzing Cultures: An Introduction and Handbook*. Indiana University Press.

Di Giovine, M. A., & Garcia-Fuentes, J. M. (2016). Sites of Pilgrimage, Sites of Heritage: An Exploratory Introduction. *International Journal of Tourism Anthropology*, 5(1–2), 1–23.

Eliade, M. (1959). *The Sacred and the Profane: The Nature of Religion*. Harcourt.

Harding, T. (2019). Heritage Churches as Post-Christian Sacred Spaces: Reflections on the Significance of Government Protection of Ecclesiastical Heritage in Swedish National and Secular Self-Identity. *Culture Unbound*, 11(2), 209–230. https://doi.org/10.3384/cu.2000.1525.20190627

Knippschild, R., & Zöllter, C. (2021). Urban Regeneration Between Cultural Heritage Preservation and Revitalization: Experiences with a Decision Support Tool in Eastern Germany. *Land (Basel)*, 10(6), 547–. https://doi.org/10.3390/land10060547

Knott, K. (2015). *The Location of Religion: A Spatial Analysis*. Routledge. https://doi.org/10.4324/9781315652641

Lisle, D. (2021). A Speculative Lexicon of Entanglement. *Millenium: Journal of International Studies*, 49(3), 435–461. https://doi.org/10.1177/030582 98211021919

Lung, D. (2012). Built Heritage in Transition: A Critique of Hong Kong's Conservation Movement and the Antiquities and Monuments Ordinance. *Hong Kong Law Journal*, 42(1), 121–141.

Olsen, D. H. (2003). Heritage, Tourism, and the Commodification of Religion. *Tourism Recreation Research*, 28(3), 99–104. https://doi.org/10.1080/025 08281.2003.11081422

Salemink, O. (2016). Described, Inscribed, Written Off: Heritagisation as (Dis)Connection. In P. Taylor (Ed.), *Connected and Disconnected in Vietnam: Remaking Social Relations in a Post-socialist Nation* (pp. 311–346). Australian National University Press.

Salemink, O., Poulsen, R. R., & Ahl, S. I. (2020). Sacred But Not Holy: Awe, Spectacle, and the Heritage Gaze in Danish Religious Heritage Contexts. *Anthropological Notebooks*, 26(3), 70. https://doi.org/10.5281/zenodo.460 4148

Smith, L. (2006). *Uses of Heritage*. Routledge.

Sofield, T. H. B., & Li, F. M. S. (2020). Crown Cave, Guilin: A Chinese Perspective on Responsible Rural Tourism. In V. Nair, A. Hamzah, & G. Musa (Eds.), *Responsible Rural Tourism in Asia* (pp. 41–60). Channel View Publications.

Sofield, T. H. B., & Sivan, A. (1994). From Cultural Festival to International Sport—The Hong Kong Dragon Boat Races. *The Journal of Sport Tourism, 1*(3), 5–17. https://doi.org/10.1080/10295399408718541

Thouki, A. (2022). Heritagization of Religious Sites: In Search of Visitor Agency and the Dialectics Underlying Heritage Planning Assemblages. *International Journal of Heritage Studies: IJHS, 28*(9), 1036–1065. https://doi.org/10.1080/13527258.2022.2122535

Tsivolas, T. (2019). The Legal Foundations of Religious Cultural Heritage Protection. *Religions (Basel, Switzerland), 10*(4), 283–. https://doi.org/10.3390/rel10040283

UNESCO. (1994). *Nara Document on Authenticity.* UNESCO.

UNESCO. (2013). *Operational Guidelines for the Implementation of the World Heritage Convention.* UNESCO.

UN-Habitat Commission. (2022). *Learn More About Us.* https://unhabitat.org/about-us/learn-more#:~:text=UN%2DHabitat%20is%20the%20United,of%20adequate%20shelter%20for%20all

van der Tol, M., & Gorski, P. (2022). Secularisation as the Fragmentation of the Sacred and of Sacred Space. *Religion, State & Society, 50*(5), 495–512. https://doi.org/10.1080/09637494.2022.2144662

Van den Hemel, E., Salemink, O., and Stengs, I. (eds.) (2022). *Managing Sacralities: Competing and Converging Claims of Religious Heritage.* Berghahn Publications.

Venturini, T. (2010). Diving in Magma: How to Explore Controversies with Actor-Network Theory. *Public Understanding of Science, 19*(3), 258–273.

Vukonic, B. (2002). Religion, Tourism and Economics: A Convenient Symbiosis. *Tourism Recreation Research, 27*(2), 59–64. https://doi.org/10.1080/02508281.2002.11081221

Yanata, K., & Sharpley, R. (2021). Coexistence Between Tourists and Monks. Managing Temple-Stay Tourism at Koyasan, Japan. In D. H. Olsen & D. Timothy (Eds.), *The Routledge Handbook of Religious and Spiritual Tourism* (pp. 398–410). Routledge.

Zhang, Y., & Wu, Z. (2016). The Reproduction of Heritage in a Chinese Village: Whose Heritage, Whose Pasts? *International Journal of Heritage Studies, 22*(3), 228–241.

CHAPTER 10

Empowerment for Sustainable Development

Abstract Community-based empowerment can occur when politics, policies and governance align to confer power to a community to pursue its own development. There will always be certain constraints under which a community must work (such as a national building standards code and business operating licences) but empowerment will be facilitated where the following five key factors are present: (i) ownership of land and related resources; (ii) the capacity by a community to take and make decisions themselves and then, importantly, implement them; (iii) a corollary of which is that there are local stakeholders within the community who possess the capacity to exercise leadership and gain support for their vision or a better future. A lack of expertise and human resources has a negative effect; (iv) ownership or at least majority control of any business or development that a community establishes; and (v) the ability to exercise control over their own values, heritage and traditions (Sofield, Empowerment for Sustainable Tourism Development, Pergamon, 2003). The development of Yim Tin Tsai as pursued by the local community over the past two decades is a graphic example of empowerment arising from the processes of heri-ligion.

Keywords Empowerment · Governance · Power · Land and related resources ownership · Chinese customary land tenure system · New Territories Ordinance

The analysis of Yim Tim Tsai through the application of our concept of heri-ligion would not be complete without considering the issue of empowerment. Power, politics, policies and governance provide the framework and set the parameters for community empowerment to be able to emerge or, alternatively, to be suppressed. If the totality of the social and geo-political environment governing politics, policies and power relationships is disabling, where institutions and legislation incapacitate rather than enable minority communities and/or lower socio-economic classes to contribute to their pursuit of development however they may define it, then disempowerment rather than empowerment will be the outcome. Communities in many countries are left outside the governance and decision-making processes with the result that policies and decisions are made *for* them not *by* them where top-down governance is the norm. In other countries there are 'consultations with communities' which amount to little more than tokenism, leaving real decision-making in the hands of the Government and/or commercial developers. And as noted in Chapter 11, in China the Government actively encourages investment from outside agencies and in general favours developers over communities (Sofield & Li, 2020; Zhang et al., 2010). We argue that without empowerment sustainable development is difficult to attain, and to be effective, empowerment for both communities and individuals within a community must transform them from being recipients of 'development' instituted by outside forces to agents of change where they may pursue their own aspirations relatively independently.

Where a community is able to maintain a degree of ownership over its heritage and traditions (its beliefs, value system and behaviour) these factors are fundamental to their identity, assisting them to define who they are both to themselves and to outsiders; they provide a sense of belonging in a cultural sense and in terms of place and thus achieve cultural empowerment. Scheyvens (1999) attributes pride in one's own heritage as psychological empowerment, and this can be both societal (shared by the whole community) and individual simultaneously.

Five key factors that facilitate sustainable community-based empowerment are: (i) a general environment of enabling conditions where there is genuine ownership of land and control over related resources; (ii) the capacity by a community to take and make decisions themselves and then, importantly, implement them; (iii) a corollary of which is that there are local stakeholders within the community who possess the capacity to exercise leadership and gain support for their vision of a better future.

A lack of technical expertise and human resources may have a negative effect but where local leadership is able to harness expertise from outside sources this lack may be mitigated; (iv) ownership or at least majority control of any business or development that a community establishes; and (v) the ability to exercise control over their own values, heritage and traditions (Sofield, 2003). Heritage and tradition are not static and unchanging, of course, but dynamic. Each generation tends to redefine its traditional heritage in response to new understandings, new experiences and new inputs from an ever-increasing range of sources, both internal and internal. The means by which the almost inconsequential period of time that Father Freinademetz spent in Yim Tin Tsai became elevated through his sainthood (due to the agency of the far distant authority of the Vatican in Rome and the instrumental role of the Vicar General of the Catholic Church in Hong Kong), is a classic example of the way in which heritage may be claimed and redefined, and can set a community on a completely novel course never previously imagined or planned—that of heritage-making through restoration of village resources where sacralization dominated secularization, leading to pilgrimage visitation. As is the case with Yim Tin Tsai, interpretation of the past (historically, materially and spiritually) may change to suit or satisfy particular needs and thus is to be viewed as a process rather than a static entity (Sofield, 2003).

Much more could be written about empowerment than this short introduction, but the issue of land ownership in the context of Yim Tin Tsai merits supplementary attention. Because the community has experienced unchallenged ownership of the island (a vital enabling condition) it has been able to implement its own development programme and this factor thus constitutes the foundation building block for its empowerment. Land ownership links into and underpins attachment to place, to home, to cultural heritage and tradition, especially with reference to ancestor worship, kinship and lineage anthologies. With reference to the latter three points, a customary widespread land tenure system in ancient China dating back more than 1000 years to the Song Dynasty and sustained by imperial governance was the institution of *t'so*, and subsequently under the Qing Dynasty in the seventeenth century, the institution of *t'ong* (Hayes & Ching, 1976; Wolf, 1974). They were also present in the Kowloon Peninsula and the New Territories where ancestors of prominent local clans settled centuries ago. The traditional custom of establishing *t'so* and *t'ong* was mainly for ancestor veneration and clan solidarity, having its origins in Daoist and Confucian respect for authority and ancestors. *T'so* is described as: "*an ancient Chinese institution of ancestral land-holding whereby land derived from a common ancestor is enjoyed by his male descendants for their lifetimes and so from generation to generation indefinitely*" (Cheung, 2022, p. 1). *T'ong*, literally translated, means "*hall*" or "*gathering place*", and as a lineage-based institution (often in the form of a secret society) they inter alia established ancestral halls, recorded clan lineages, educated male youths and promulgated the Daoist belief system (Cheung, 2022; Chiang, 2023; Ebrey, 1999). While *t'so* and *t'ong* began to wane in China in the nineteenth century, and completely disappeared under Mao's 1949 land reforms of course, they survived in Hong Kong and the New Territories because of the British colonial policy of accepting some Chinese customs and customary rights that were written into colonial law.

Hong Kong Island was ceded to the British Government by the Qing Dynasty Government in 1842 and the colony became Crown Land with plots leased on 75- or 99-year terms: only one freehold title was granted, that of the Anglican cathedral of St John's in the city centre on 4000 sq metres (Caudevilla, 2016). Concerned that the Island's security could be compromised if hostile sources were to gain control of the Kowloon Peninsula and its mountain range that overlooked Victoria Harbour, the British Government negotiated a 99-year lease of the New Territories with the Qing Government (Convention of Peking, 1898). It stretched from Kowloon to the Sham Chun (Shenzhen) River separating the peninsula from mainland China and covered 376 sq miles (974 square kilometres). It included 235 islands, of which Yim Tin Tsai was one. A survey commissioned by the British Administration of Hong Kong at the time indicated that there were 423 villages with a total population of 100,320. Of these, 255 were Hakka villages with a population of 36,020 (Industrial History of Hong Kong Group, 2014). They were mainly subsistence rice farmers and fishing communities who held tenure under the long-established Chinese customary land rights of *T'so/T'ong* (Caudevilla, 2016; Chiang, 2023).

On securing the lease from China, the British legal and land lease systems were extended from Hong Kong Island to the New Territories, sparking an armed insurrection in 1899 ("The Six Day War") by Chinese villagers fearful that they would lose their lands. As Hase (2013, p. 2) stated: "*For the villagers of the traditional society of the New Territories of Hong Kong, there was nothing more important than their ownership and control of rice-land"*. After suppressing the rebellion, the British colonial regime passed the 1900 New Territories Ordinance which enshrined Chinese customs and customary rights in relation to land in the New Territories but changed traditional 'in perpetuity' ownership to the colonial lease system (Hase, 2008). The delineation and formalization of *t'so* and *t'ong* found expression in Sect. 13 of the New Territories Ordinance which conferred on the Supreme Court of Hong Kong and the District Courts the jurisdiction: *"to recognize and enforce any Chinese custom or customary right affecting [New Territories] land'. The tso and the family tong are lineage trusts whereby land is held in common ownership for the benefit of the whole lineage"* (Hong Kong Government, 1900). To further allay concerns of a Government take-over of their lands, the colonial regime identified the traditional owners in every village and they were granted 'Crown Block Leases' (i.e. not as individual titles but as collective ownership under lineage trusts) most of them for 99 years until 1997 when the lease of the New Territories would expire (Caudevilla, 2016; Chiang, 2023; Wong, 1990). By this recognition of ancestral land rights the Chan clan of Yim Tin Tsai and Kau Sai Chau were granted control of their islands under the British colony of Hong Kong.

With the hand-over of Hong Kong and the New Territories to the Chinese Government in 1997, the Sino-British Joint Declaration treaty set the conditions of transfer whereby China agreed to maintain existing structures of Government and economy under the principle of "One country, two systems" for a period of 50 years after 1997. The colonial-initiated leasehold system of ancestral land rights stayed broadly unchanged after reunification with the Mainland, with many customary-based leases extended for those 50 years to 2047, even though the traditional land tenure system was incompatible with the mainland's system.

Following the Introduction of revised legislation in Hong Kong in 2017 that in fact introduced significant changes to the governance system and authority of the Hong Kong Special Administrative Region Government to align it more closely with mainland China, the Hong Kong Lands Council issued a statement affirming that all leases would remain valid and that those leases which were due to expire in 2047 *could* be renewed for 50 years (Legislative Council of Hong Kong, 2017). Thus, despite the various legislative and other changes arising from the 1997 reunification of Hong Kong and the New Territories with China, and the lack of certainty after 2047, for the past two decades and for the next two decades the Yim Tin Tsai community, in association with the Catholic community of Hong Kong, have been and will be able to proceed with their plans for rejuvenation of the island through heritage-making unhindered to a significant extent by Government legislation, bureaucracy and officials' ingression into their affairs. The fact that several former residents have applied to build new houses under the 2016 Act on Small House Policy and Planning for Indigenous Villages (revised from the original Act of 1972, (referenced in Chapters 9 and 11) is testament to the villagers' confidence in continuing to pursue their aspirations for a rejuvenated Yim Tin Tsai under the land tenure system. The New Territories Ordinance and subsequent acceptance of its provisions related to land tenure in the New Territories such as the Small House Policy (which by legal definition includes Yin Tin Tsai) over the past 120 years by successive governments and administrations have been fundamental in creating that enabling environment necessary for empowerment.

As noted in this chapter on stakeholder analysis, the roles and the ascribed status of the Catholic Vicar General of Hong Kong, Father Chan, and of the Village Council chief, Colin Chan, accorded them instrumental authority and empowered them individually, so that their effective leadership may also be identified as a key factor in empowering the community. The various heritage-making initiatives, both sacred and secular that commenced in 2003 with the restoration of St Joseph's chapel, the development of the Trail of Reconciliation, the subsequent restoration of the salt pans and transformation of the former primary school into the Hakka Cultural Exhibition Centre, accompanied by burgeoning pilgrimage and secular tourism with necessary support services and facilities, demonstrate the principles of sustainability through enduring community support. In this context, the social bonds between the Yim Tin Tsai community and the Catholic Church of Hong Kong constitute another important factor for enabling empowerment, i.e. the presence/existence of 'social capital' both within and between the island community's families where the connections are organic, and 'bridging social capital' as defined by Marcinek and Hunt (2015), across to the Catholic Church.

Yim Tin Tsai thus represents a clear case of sustainable community empowerment and it may be attributed largely to those five factors noted above: an enabling environment where there is unchallenged ownership of land and accompanying resources, ownership of resulting economic/

business activities, the capacity to make and implement their own decisions, effective stakeholder leadership and human resources within the community and custodial command over its heritage and traditions.

REFERENCES

Caudevilla, O. (2016). The System of Landholding in Hong Kong. *Inter Asia Papers, 52*(1), 1–14.

Cheung, Chi-Fai. (2022). *Tso and Tong in the New Territories*. Hong Kong: Legislative Council Secretariat of the Hong Kong Special Administrative Region. https://www.legco.gov.hk/research-publications/english/essentials-2022ise08-tso-and-tong-in-the-new-territories.htm

Chiang, C. K. (2023). Beyond Legal Pluralism: Chinese Customs and Customary Laws in Colonial Hong Kong (1841–1997). In *Translocal Chinese: East Asian Perspectives*, On-Line publication by Brill. https://brill.com/view/journals/tcea/17/1/article-p58_004.xml?language=en#mainContent

Ebrey, P. B. (1999). *The Cambridge Illustrated History of China*. Cambridge University Press.

Hase, P. H. (2013). *The Six-Day War of 1899: Hong Kong in the Age of Imperialism*. Royal Asiatic Society Hong Kong Studies Series.

Hase, P. H., & Yan, S. (2009). The Price and Consumption of Salt in China in 1901. *Journal of the Royal Asiatic Society Hong Kong Branch, 49*, 127–218.

Hayes, J., & Ching, S. W.-t. (1976). *Rural Society and Economy in Late Ch'ing: A Case Study of the New Territories of Hong Kong*. https://muse.jhu.edu/article/398564/pdf

Hong Kong Government (1900, Rev. 1984). *New Territories Ordinance*. Hong Kong.

Industrial History of Hong Kong Group (2014). *New Territories Population—1898*. https://industrialhistoryhk.org/new-territories-population-1898/. Accessed 20 January 2024.

Marcinek, A. A., & Hunt, C. A. (2015). Social Capital, Ecotourism, and Empowerment in Shiripuno, Ecuador. *International Journal of Tourism Anthropology, 4*, 327 [Google Scholar].

Scheyvens, R. (1999). Ecotourism and the Empowerment of Local Communities. *Tourism Management, 20*, 245–249.

Sofield, T. H. B., & Li, F. M. S. (2020). Crown Cave, Guilin: A Chinese Perspective on Responsible Rural Tourism. In V. Nair, A. Hamzah, & G. Musa (Eds.), *Responsible Rural Tourism in Asia* (pp. 41–60). Channel View Publications.

Sofield, T. H. B. (2003). *Empowerment for Sustainable Tourism Development*. Pergamon.

Wolf, A. P. (Ed.). (1974). *Religion and Ritual in Chinese Society*. Stanford University Press.

Wong, B. (1990). Chinese Customary Law—An Examination of T'sos and Family T'ongs. *Hong Kong Law Journal, 20*, 13–30.

Zhang, ChaoZhi, Fyall, A. and Zheng, Yenfen (2010). Conflict within World Heritage Sites in China: A Longitudinal Study. *Current Issues In Tourism, 18*(2): 110–116 https://doi.org/10.1080/13683500.2014.912204

CHAPTER 11

Future Research and Conclusions

Abstract Areas for future research that could prove fruitful in enhancing our concept of heri-ligion may be its application to heritage sites of other religions such as Islam, Hinduism, Buddhism and so forth; environmental impacts of religious tourism and mass pilgrimage (a field that is quite under-studied); and perceptions of museums where they reframe secular and sacred artefacts so that their interpretation and presentation may create contestation and entanglement.

Keywords Application of heri-ligion to other religious sites · Islam · Hinduism · Busshism · Environmental impacts of religious tourism · Museum presentation and interpretation of sacred and secular artefacts

As with any research there are always new areas to explore and we consider that it could prove fruitful in enhancing our concept of heri-ligion to extend its application to heritage sites of other religions such as Islam, Hinduism, Buddhism and so forth; environmental impacts of religious tourism and mass pilgrimage (a field that is quite under-studied); and perceptions of museums where they reframe secular and sacred artefacts so that their interpretation and presentation creates contestation and entanglement. The concept of empowerment as an outcome of the processes of heri-ligion could also be considered in greater detail.

© The Author(s), under exclusive license to Springer Nature   121
Singapore Pte Ltd. 2024
T. Sofield et al., *Heritage-Making in Hong Kong Through Culture and Religion*, https://doi.org/10.1007/978-981 97-4339-1_11

In addition to Christian places such as Yim Tin Tsai it would be useful to consider other religions and sites where tourism and secular heritage values have contributed to site transformation. They could include *inter alia* Turkiye's and Iran's Islamic world heritage sites, Nepal's Buddhist and Hindu world heritage sites, Japan's Shinto heritage sites and India's many Hindu heritage sites, potentially attesting to the validity of our concept or possibly unearthing new elements to enrich the construct.

One area of potential research interest that we have not pursued focuses on the environmental impacts that have arisen from the heratigization of Yim Tin Tsai. Over 20 years this small island has experienced exponential growth in visitor numbers, from a very small number of curious 'passers-by' prior to 2003 to more than 100,000 visitors in 2023. We have no baseline studies of key indicators that allow us to measure what anthropocentric impacts and degradation would have resulted from this transformation of an abandoned environment to one experiencing relatively massive human ingression, although logic suggests it could be substantial. Shinde and Olsen (2023) have challenged existing perceptions of pilgrimage tourism as environmentally sound in an incisive article that questions its sustainability, drawing attention to an inherent contradiction between the pollution caused by mass pilgrimage to sites such as Mecca in Saudi Arabia, Varansi in India and Fatima in Portugal with tenets of their scripture and teachings that promote stewardship and sustainable interactions with the natural and built environment. Visitation in the millions causes *"congestion, vandalism, noise pollution, microclimatic change … damage to fauna and flora, particularly when religious sites are located inside protected areas, increased air and water pollution, deforestation … offerings brought by pilgrims for ritual purposes that are often left to rot …",* problems which the behaviour of many pilgrims exacerbate and ignore and that are left for local authorities and host communities with often inadequate waste management and sewage systems to handle (Shinde & Olsen, p. 464). Sofield (2001) encountered similar issues with pilgrimage tourism to the World Heritage Sites of the Kathmandu Valley, Nepal—Badrinath and Swayambhu(both Buddhist), and Changu Narayan and Pashupati (both Hindu)—where the four Nepalese host communities responsible for the custodianship of the temples and stuppas were highly critical of the behaviour of pilgrims. The sites showed evidence of all of the environmental impacts listed by Shinde and Olsen above, with the host communities surprisingly stating that they preferred secular tourists and westerners because they respected the rules, used the toilets and waste

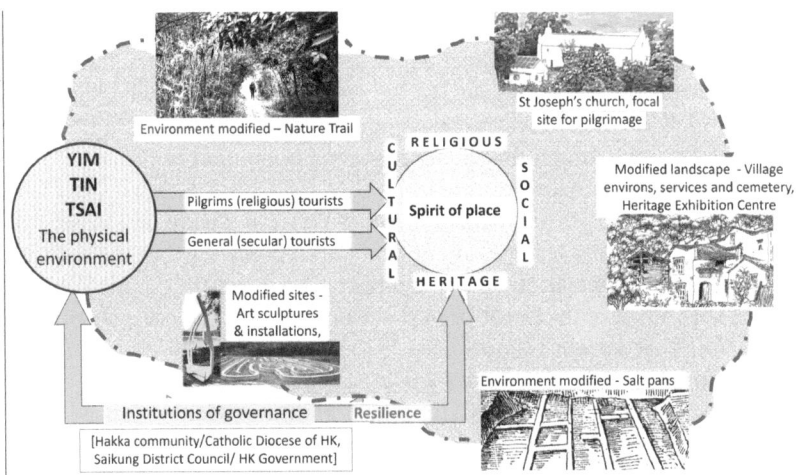

Diagram 11.1   A conceptual model of the relationships between the environment, religion and tourism in Yim Tin Tsai (adapted from Shinde & Olsen, 2023)

disposal bins, did not camp in the sacred hillside forests around each site where pilgrims degraded the environment by cutting trees for fuel and frightening birds and fauna away, excreting human waste and leaving rubbish that the locals had to clean up. A specific contrast was also drawn between social behaviour: pilgrims, the custodians said, treated their villages as if they owned them and invaded their privacy, even entering their houses uninvited: secular tourists on the other hand respected their privacy and were amenable to directions. Shinde and Olsen (2023) have developed a model to assess environmental issues which highlights the importance of institutions in pilgrim-town governance. They argue that resilience of the involved institutions in managing their responsibilities is fundamental to ensure sustainability. Diagram 11.1 adapts their model to illustrate the relationship between the environment, religion and tourism in Yim Tin Tsai.

With reference to Yim Tin Tsai there is a trialectic relationship that rests with the Hakka community, its alliance through the Joint Organizing Committee with the Catholic Diocese of Hong Kong and the Sai Kung District Council/Government of Hong Kong. The first holds primary responsibility as the landowner for mediating the physicality of the sacred and secular heritage of the island with tourists; the second holds primary responsibility for mediating the Christian spirituality of the place; and through their pilgrimage/visitor management body, "Salt and Light" together with Government agencies that oversee heritage policy and provide supporting infrastructure, share a combined responsibility for environmental sustainability.

It is considered that the management of Yim Tin Tsai would benefit by an environmental impact assessment that could contribute to the forward planning of the destination as its visitation continues to expand. We thus suggest that an eleventh factor be added to our composite concept of heri-ligion, and that is: Heri-ligion: (xi). *Environmental impacts of religious heritagization: sustainability and regeneration.*

Concerning Yin Tim Tsai's Heritage Exhibition Centre (an example of adaptive re-use with the primary school now functioning as a folk museum), Berns (2015, p. 1) suggests that *the study of religious dimensions of visitor experiences in public museums is under-researched,* in part because museums in general are perceived as spaces that are secular in nature, that *material religious exhibits are displayed in public spaces not intended to be devotional,* and thus *the processes through which visitors may experience sacred presences in a museum* are rarely analysed. In Yim Tin Tsai's folk museum, visitors experience embodied interactions with elements of the sacred that are intermingled with secular objects. Bern's findings, applying Actor Network Theory (Latour, 2004) to observe visitor behaviour in the British Museum in London, "*revealed how perceptions of the museum as secular, shaped visitor norms and thereby influenced whether the museum became a site of conflict or opportunity for sacred encounters* (2015, p. 2). Utilizing Beren's approach with our concept of heri-ligion, applying relevant components in a study of the Yim Tin Tsai Heritage Exhibition Centre, would realize a constructive area for future research.

The concept of heri-ligion, we suggest, provides a useful new tool for analysing the dynamic characteristics of a site that is experiencing sacralization/de-sacralization/heritagization, moving beyond the normative association of tension connotative of entanglement to a wider theoretical framework to unpack also adaptations that are harmonious and/or tolerant, as exist in instances of syncretism, symbiosis, mutualism and commensalism. We argue that stakeholder analysis is fundamental for determining the drivers behind the who, how and why of change processes and thus has an integral place in our concept of heri-ligion. As graphically illustrated by our application of heri-ligion to Yim Tin Tsai, for its community a strong sense of attachment to place, allied with a consciousness of 'home' and awareness of traditions, heritage and cultural roots, are pivotal structures that open windows of discovery into the subtleties and undercurrents of the transformative forces at work. While the island's rejuvenation was originally based on its Catholic background with restoration of the church as the focal point for almost exclusively pilgrimage visitation, the gradual secularization of sites and other attractions such as the restoration of the salt pans, the de-sacralization of the Trail of Reconciliation to a dual purpose as a Nature Trail, and the establishment of the Hakka Traditional Exhibition Centre as a folk museum in the re-purposed school house (albeit with numerous religious artefacts and photographs), have led to a gradual expansion of general tourism where entanglement has only lately become a more relevant research tool in deciphering the change agents underlying Yim Tin Tsai's rejuvenation from an abandoned village of dilapidated ruins to a destination of some significance in Hong Kong. We would also suggest however, that different types of tourists, with quite different motivations for visiting the island, often move seamlessly between the 'sacred' and the 'secular' with appreciation of the destination's varied sites/sights and range of experiences they engender, in which spirituality is not uncommonly generated, reinforcing our view that a sacred/secular gaze is extant for many (perhaps even most) visitors. In 2020, Collins Kreiner commented on the fragmented nature of academic studies in the field of religion and tourism and suggested there was a need for a holistic approach and new avenues of research: in this context we offer our paper as one contribution to address the issue. We thus present the multiple components of our concept of heri-ligion as a useful tool for researching the dynamics of development of destinations where religion and heritage co-mingle, especially for those sites and situations where local communities engaging in co-creation have played an active role in determining outcomes.

The net result of what has been happening on Yim Tin Tsai over the past two decades has been a graphic example of community-led empowerment for sustainable development. Politics, policies and governance have all aligned to confer a degree of power on the community of Yim Tin Tsai and the Catholic Church of Hong Kong so that, acting in partnership they have been able to make their own decisions more or less independently and set their own agenda for advancing their interests as self-defined. Heritage-making and religious tourism now extended to secular tourism have been the tools that the community have used, and they have been successful through community-based empowerment.

REFERENCES

Berns, S. (2015). *Sacred Entanglements: Studying Interactions Between Visitors, Objects and Religion in the Museum* [Unpublished Doctoral Dissertation, University of Kent]. https://kar.kent.ac.uk/50505/1/202StephBerns_PhDThesis.pdf

Latour, B. (2005). *Reassembling the Social: An Introduction to Actor-Network-Theory*. Oxford University Press.

Shinde, K. A., & Olsen, D. H. (2023). Reframing the Intersections of Pilgrimage, Religious Tourism, and Sustainability. *Sustainability (Basel, Switzerland)*, *15*(1), 461–. https://doi.org/10.3390/su15010461

Sofield, T. H. B. (2001). Sustainability and Pilgrimage Tourism in the Kathmandu Valley of Nepal. In V. Smith & Brent, M. (Eds), *Hosts and Guests Re-visited: Tourism Issues of the 21st Century* (Ch. 20, pp. 257–274). Cognizant Communications Corp.

REFERENCES

Ahrentzen, S. B. (2012). Home as a Workplace in the Lives of Women. In S. M. Low & I. Altman (Eds.), *Place Attachment: A Conceptual Inquiry* (Ch. 6, pp. 113–123). New York & London: Plenum Press.

An, J., & Wang, F. (2011). Consulting Management Offered by Hotel Management Companies: The Exploration and Practices of Folk Culture Villages. In G. Zhang, R. Son, & D. Liu (Eds.), *Green Book on China's Tourism 2011: China's Tourism Development Analysis and Forecast (English Version edited by A. Wolfgang)* (pp. 215–224). Social Sciences Academic Press.

Ap, J., Li, F. M. S., & Sofield, T. H. B. (1994). *The Saikung Region of Hong Kong: A Tourism Strategy*. Saikung District Council.

Ashtari, H. (2022, November 4). *What Are Haptics? Meaning, Types, and Importance*. Spiceworks. https://www.spiceworks.com/tech/tech-general/articles/what-are-haptics/

Ashworth, G. J. (2000). Heritage, Tourism and Places: A Review. *Tourism Recreation Research, 25*(1), 19–29. https://doi.org/10.1080/02508281.2000.11014897

Ashworth, G. J. (2011). Preservation, Conservation and Heritage: Approaches to the Past in the Present Through the Built Environment. *Asian Anthropology, 10*(1), 1–18. https://doi.org/10.1080/1683478X.2011.10552601

Ashworth, G. J., & Graham, B. (Eds.). (2017). *Senses of Place: Senses of Time*. Routledge.

Astor, A., Burchardt, M., & Griera, M. (2017). The Politics of Religious Heritage: Framing Claims to Religion as Culture in Spain. *Journal for the Scientific Study of Religion, 56*(1), 126–142. https://doi.org/10.1111/jssr.12321

Atha, M. (2014). *Archaeological Survey-Cum-Excavation, Yim Tin Tsai, Sai Kung (Oct.–Nov. 2013)*. Hong Kong Archaeological Society.

Aulitzky, J. M. (1932). *Kurzes Lebensbild des P. J. Freinademetz*. Mödling

Bai, Y. (2003). *On the Early City and the Beginning of the State in Ancient China*. Bureau of International Cooperation, Hongkong, Macao and Taiwan Academic Affairs Office. Chinese Academy of Social Sciences. http://www.worldlibrary.org/Articles/AncientChineseurbanplanning=Bai,Yunxiang

Balkenhol, M., van den Hemel, E., & Stengs, I. (2020). Introduction: Emotional Entanglements of Sacrality and Secularity—Engaging the Paradox. In M. Balkenhol, E. van den Hemel, & I. Stengs (Eds.), *The Secular Sacred*. Palgrave Macmillan.

Bauman, Z. (1996). From Pilgrim to Tourist—Or a Short History of Identity. In S. Hall & P. Gay (Eds.), *Questions of Cultural Identity* (pp. 18–36). Sage.

Bauman, Z. (2001). *The Individualized Society*. Polity Press.

Baur, J. (1939). *Der Diener Gottes P. Joseph Freinademetz SVD. 1852–1908. Das Leben eines heiligmäßigen Chinamissionärs dem Volke erzählt, Missionari Verbiti*. Varone.

Berns, S. (2015). *Sacred Entanglements: Studying Interactions Between Visitors, Objects and Religion in the Museum* [Unpublished Doctoral Dissertation, University of Kent]. https://kar.kent.ac.uk/50505/1/202StephBerns_PhDThesis.pdf

Bhabha, H. K. (1994). *The Location of Culture*. Routledge.

Bogumil, Z., & Lukaszewicz, M. (2018). Between History and Religion: The New Russian Martyrdom as an Invented Tradition. *East European Politics and Societies, 32*(4), 936–963. https://doi.org/10.1177/0888325417747969

Bourdieu, P. (1993). *The Field of Cultural Production: Essays on Art and Literature*. New York: Columbia University Press.

Bornemann, F. (1926). *As Wine Poured Out. Blessed Joseph Freinademetz. Missionary in China 1879–1908* (Fr. J. Vogelgesang, Trans., 1984). Divine Word Missionaries.

Boym, S. (2002). *The Future of Nostalgia*. Basic Books.

Bremer, T. S. (2006). Sacred spaces and tourist places. In D. J. Timothy & D. H. Olsen (Eds.), *Tourism, religion and spiritual journeys*. (pp.25-35). Routledge.

Brown, B., & Perkins, D. (2012). Disruptions in Place Attachment. In S. M. Low & I. Altman (Eds.), *Place Attachment: A Conceptual Inquiry* (ch.13, pp. 279–299). New York & London: Plenum Press.

Casanova, J. (2009). The Secular and Secularisms. *Social Research, 76*(4), 1049–1066. https://doi.org/10.1353/sor.2009.0064

Catholic Heritage Organization. (2016). *St Joseph's Chapel, Yim Tin Tsai*. https://www.catholicheritage.org.hk/en/catholic_building/yim_tin_tsai/index.html

Caudevilla, O. (2016). The System of Landholding in Hong Kong. *Inter Asia Papers, 52*(1), 1–14.

Central Committee of the Chinese Communist Party. (2005). *New Socialist Countryside Policy.* CCP.

Chambers, R. (1997). *Whose Reality Counts? Putting the First Last.* Intermediate Technology Publications.

Chamaz, Kathy (2006). *Constructing Grounded Theory. A Practical Guide Through Qualitative Analysis.* London: Sage Publications

Chan, D., Cheung, L., & Chan, L. Y. (2012). *The Catholic Nature Trail of Reconciliation.* Caritas Printing Centre.

Chan, E., Lam, T., Shiu, D., & Wong, M. (2014). *Renaissance in Yim Tin Tsai: Catholic Traditions and Salt-Making History define Hakka Village—Varsity.* Chinese University of Hong Kong.

Chang, C.-C., & Lin, Y.-H. (2022). Constructing Hakka Ethnic Identity Through Narrative Genealogy Writing. *SAGE Open, 12*(1). https://doi.org/10.1177/21582440221079913

Chapman, W. (Ed.). (2020). *Asia Conserved, vol. IV: Lessons Learned from the UNESCO Asia-Pacific Heritage Awards for Culture Heritage Conservation, 2015–2019.* Southeast University Press. https://unesdoc.unesco.org/ark:/48223/pf0000374413

Charmaz, K. (2014). *Constructing Grounded Theory: A Practical Guide Through Qualitative Analysis* (2nd ed.). Sage.

Cheung, Chi-Fai. (2022). *Tso and Tong in the New Territories.* Hong Kong: Legislative Council Secretariat of the Hong Kong Special Administrative Region. https://www.legco.gov.hk/research-publications/english/essentials-2022ise08-tso-and-tong-in-the-new-territories.htm

Chiang, C. K. (2023). Beyond Legal Pluralism: Chinese Customs and Customary Laws in Colonial Hong Kong (1841–1997). In *Translocal Chinese: East Asian Perspectives*, On-Line publication by Brill. https://brill.com/view/journals/tcea/17/1/article-p58_004.xml?language=en#mainContent

Ching, J. (1993). *Chinese Religions.* Orbis Books.

Clark, B. (2016). Community Narratives. In L. A. Jason & D. S. Glenwick (Eds.), *Handbook of Methodological Approaches to Community-Based Research: Qualitative, Quantitative, and Mixed Methods* (pp. 43–52). Oxford University Press.

Choon, Y. N. (2005). *The Hakka Chinese: Their Origin, Folk Songs and Nursery Rhymes.* Poseidon Books.

Chu, W. H. (1998). *Conservation of Terrestrial Biodiversity in Hong Kong.* Unpublished M.Phil. thesis, The University of Hong Kong.

Collins Kreiner, N. (2020). Religion and Tourism: A Diverse and Fragmented Field in Need of a Holistic Agenda. *Annals of Tourism Research, 82*(2). https://doi.org/10.1016/j.annals.2020.102892

Constable, N. (1994). *Christian Souls and Chinese Spirits. A Hakka Community in Hong Kong.* University of California Press.

Constable, N. (2005). *Guest People: Hakka Identity in China and Abroad.* University of Washington Press.

Cross, F. L., & Livingstone, E. A. (2005). *The Oxford Dictionary of the Christian Church* (3rd ed.). Oxford University Press.

Danesi, M., & Perron, P. (1999). *Analyzing Cultures: An Introduction and Handbook.* Indiana University Press.

de Jong, F., & Mapril, J. (Eds.). (2023). *The Future of Religious Heritage: Entangled Temporalities of the Sacred and the Secular.* Routledge.

Deschepper, J. (2018). Le «patrimoine soviétique» de l'URSS à la Russie contemporaine Généalogie d'un concept. *Vingtième siècle (Paris. 1984), 137*(1), 77–98. https://doi.org/10.3917/ving.137.0077

Desvallees, A., & Mairesse, F. (Eds.). (2009). *Key Concepts of Museology.* International Council of Museums, Armand Collin.

Di Giovine, M. A. (2012). Padre Pio for sale: Souvenirs, Relics, or Identity Markers? *International Journal of Tourism Anthropology, 2*(2), 108–127.

Di Giovine, M. A. (2021). Religious and Spiritual World Heritage Sites. In M. A. Olsen & D. Timothy (Eds.), *The Routledge Handbook of Religious and Spiritual Tourism* (ch.15). Routledge.

Di Giovine, M. A., & Garcia-Fuentes, J. M. (2016). Sites of Pilgrimage, Sites of Heritage: An Exploratory Introduction. *International Journal of Tourism Anthropology, 5*(1–2), 1–23.

Di Giovine, M. A., & Choe, J. (2019). Geographies of Religion and Spirituality: Pilgrimage Beyond the "Officially" Sacred. *Tourism Geographies, 21*(3), 361–383. https://doi.org/10.1080/14616688.2019.1625072

Douglas, A. (1994). *Symbiotic Interactions.* Oxford University Press.

Douglas, A. E. (2010). *The Symbiotic Habit.* Princeton University Press.

Easthope, H. (2009). Fixed Identities in a Mobile World? The Relationship Between Mobility, Place and Identity. *Identities: Global studies in power and culture, 16*(1), 61–82.

Ebrey, P. B. (1999). *The Cambridge Illustrated History of China.* Cambridge University Press.

Edgley, A., Stickley, T., Timmons, S., & Meal, A. (2016). Critical Realist Review: Exploring the Real, Beyond the Empirical. *Journal of Further and Higher Education, 40*(3), 316–330. https://doi.org/10.1080/0309877X.2014.953458

Eliade, M. (1959). *The Sacred and the Profane: The Nature of Religion.* Harcourt.

Encyclopaedia Britannica. (2013, February 26). The Editors of Encyclopaedia. "Hakka". *Encyclopedia Britannica.* https://www.britannica.com/topic/Hakka. Accessed 4 March 2019.

Erbaugh, M. S. (1992). The Secret History of the Hakkas: The Chinese Revolution as a Hakka Enterprise. *The Chinese Quarterly, 132*, 937–968.

European Parliament. (2015, June 24). *Report Towards an Integrated Approach to Cultural Heritage for Europe*. Committee on Culture and Education http://www.europarl.europa.eu/doceo/document/A-8-2015-0207_EN.html?redirect

Farias, M., Coleman, T. J., Bartlett, J. E., Oviedo, L., Soares, P., Santos, T., & Bas, M. del C. (2019). Atheists on the Santiago Way: Examining motivations to go on pilgrimage. *Sociology of Religion, 80*(1), 28–44. https://doi.org/10.1093/socrel/sry019

Fischer, H. (1936). *Joseph Freinademetz. Steyler Missionary in China. Ein Lebensbild (A Biography)*. Missionsdruckerei (Missions Printing Office).

Foy, G. (2023). *Chinese Religions and Philosophies: From Past to Present and Present to Past*. Asia Society. https://asiasociety.org/chinese-religions-and-philosophies

Freeman, R. E. (1984). *Strategic Management: A Stakeholder Approach*. Pitman.

Freeman, R. E. (2010). *Stakeholder Theory: The State of the Art*. Cambridge University Press.

Fuentenebro de Diego F., & Valiente Ots, C. (2014). Nostalgia: A Conceptual History. *History of Psychiatry, 25*(4), 404–411. https://doi.org/10.1177/0957154X14545290.

Giuliani, M. V. (2003). Theory of Attachment and Place Attachment. In M. Bonnes, T. Lee, & M. Bonaiuto (Eds.), *Psychological Theories for Environmental Issues* (pp. 137–170). Ashgate.

Godley, M. R. (1989). The Sojourner: Returned Overseas Chinese in the People's Republic of China. *Pacific Affairs, 62*(3), 330–352.

Graburn, N. H. H. (1977). Tourism: The Sacred Journey. In V. Smith (Ed.), *Hosts and Guests: The Anthropology of Tourism* (pp. 17–31). University of Philadelphia Press.

Graburn, N. (2001). Secular Ritual: A General Theory of Tourism. In V. Smith & M. Brent (Eds.), *Hosts and Guests Revisited: Tourism Issues of the 21st Century*. Cognizant Communications.

Graham, B., Ashworth, G. J., & Tunbridge, J. E. (2000). *A Geography of Heritage*. Arnold.

Griffin, K., & Raj, R. (2017). The Importance of Religious Tourism and Pilgrimage: Reflecting on Definitions, Motives and Data. *International Journal of Religious Tourism and Pilgrimage, 5*(3). https://doi.org/10.21427/D7242Z

Grimble, R., & Wellard, K. (1997). Stakeholder Methodologies in Natural Resource Management. A Review of Principles, Contexts, Experiences and Opportunities. *Agricultural Systems Journal, 55*(2), 173–193.

Hall, S. (1997). The Work of Representation. In S. Hall (Ed.), *Representation: Cultural Representations and Signifying Practices* (pp. 13–74). The Open University & Sage.

Harding, T. (2019). Heritage Churches as Post-Christian Sacred Spaces: Reflections on the Significance of Government Protection of Ecclesiastical Heritage in Swedish National and Secular Self-Identity. *Culture Unbound, 11*(2), 209–230. https://doi.org/10.3384/cu.2000.1525.20190627

Hardy, A. L. (2001). *A Troubled Paradise: Stakeholder Perceptions of Tourism in the Daintree Region of Far North Queensland, Australia* [Doctoral Dissertation, University of Queensland].

Harvey, D. C. (2001). Heritage Pasts and Heritage Presents: Temporality, Meaning and the Scope of Heritage Studies. *International Journal of Heritage Studies: IJHS, 7*(4), 319–338. https://doi.org/10.1080/135816 50120105534

Harvey, D. C. (2015). Heritage and Scale: Settings, Boundaries and Relations. *International Journal of Heritage Studies: IJHS, 21*(6), 577–593. https://doi.org/10.1080/13527258.2014.955812

Hase, P. H. (2013). *The Six-Day War of 1899: Hong Kong in the Age of Imperialism*. Royal Asiatic Society Hong Kong Studies Series.

Hase, P. H., & Yan, S. (2009). The Price and Consumption of Salt in China in 1901. *Journal of the Royal Asiatic Society Hong Kong Branch, 49,* 127–218.

Hayes, J., & Ching, S. W.-t. (1976). *Rural Society and Economy in Late Ch'ing: A Case Study of the New Territories of Hong Kong*. https://muse.jhu.edu/article/398564/pdf

Henninghaus A. (1920). *P. Jos. Freinademetz S.V.D. Sein Leben und Wirken. Zugleich Beiträge zur Geschichte der Mission in Süd-Schantung*. Verlag der Katholischen Mission.

HERA (Humanities in the European Research Area). (2020). *Hereligion. Heritagization of Religion and Sacralization of Heritage in Contemporary Europe*. http://heritagization.eu/ebook/

Hernández, B., Hidalgo, C., Salazar-Laplace, M., & Hess, S. (2007). Place Attachment and Place Identity. *Journal of Environmental Psychology, 27*(4), 310–319.

Hirsch, F. (2014). Empire of Nations: Ethnographic Knowledge and the Making of the Soviet Union. *Cornell University Press*. https://doi.org/10.7591/978 0801455940

Hodder, I. (2023). *Entangled: A New Archaeology of the Relationships Between Humans and Things*. Wiley.

Hollweck, S., & Ueblackner, S. (2008). *Pioneer of the Divine Word Missionaries in China—Joseph Freinademetz Serving the People of China*. https://www.svd curia.org/public/histtrad/founders/jf/jfen.htm

Home Affairs Department, Government of Hong Kong Special Administrative Region. (2023). *Rural Representative Elections.* https://www.had.gov.hk/rre/eng/index.htm

Hong Kong Government (1900, Rev. 1984). *New Territories Ordinance.* Hong Kong.

Hong Kong Intangible Cultural Office, Government of Hong Kong Special Administrative Region. (n.d.). *Hakka Unicorn Dance in Hang Hau in Saikung.* https://www.icho.hk/en/web/icho/representative_list_unicorn.html

Horii, M. (2018). *The Category of 'Religion' in Contemporary Japan: Shūkyō and Temple Buddhism* (1st ed.). Springer International Publishing.

Hsiao, H.-H. M., Peycam, P., Hui Y.-F., & Wang, S.-L. (Eds.). (2020). *Heritage as Aid and Diplomacy in Asia.* ISEAS Publishing.

Huang, G. (2021, March 3). *Yim Tin Tsai: Hong Kong's Last Salt Village.* Goldthread. https://www.goldthread2.com/travel/hong-kong-yim-tin-tsai-salt-flat/article/3126605

Hui, Y.-F., Hsiao, H.-H. M., & Peycam, P. (2018). Introduction: Finding the Grain of Heritage Politics. In H. Hsiao, Y. Hui, & P. Peycam (Eds.), *Citizens, Civil Society and Heritage-Making in Asia* (Ch.1). ISEAS Publishing.

Industrial History of Hong Kong Group. (2014). *New Territories Population—1898.* https://industrialhistoryhk.org/new-territories-population-1898/. Accessed 20 January 2024.

Isnart, C., & Cerezales, N. (Eds.). (2020). *The Religious Heritage Complex. Legacy, Conservation and Christianity.* Bloomsbury Academic.

Kirshenblatt-Gimblett, B. (1998). *Destination Culture: Tourism, Museums, and Heritage.* University of California Press.

Knapp, R. G. (1986). *Chinese Landscapes: The Village as Place.* University of Hawaii Press.

Knippschild, R., & Zöllter, C. (2021). Urban Regeneration Between Cultural Heritage Preservation and Revitalization: Experiences with a Decision Support Tool in Eastern Germany. *Land (Basel), 10*(6), 547–. https://doi.org/10.3390/land10060547

Knott, K. (2014). *The Location of Religion: A Spatial Analysis.* Routledge. https://doi.org/10.4324/9781315652641

Koltko-Rivera, M. E. (2006). Rediscovering the Later Version of Maslow's Hierarchy of Needs: Self-Transcendence and Opportunities for Theory, Research, and Unification. *Review of General Psychology, 10*(4), 302–317. https://doi.org/10.1037/1089-2680.10.4.302

Kosari, M., & Amoori, A. (2018). Thirdspace: The Trialectics of the Real, Virtual and Blended Spaces. *Journal of Cyberspace Studies, 2*(2), 163–185.

Lau, Y. C. (2010). *Estimate of Hakka Residents of Hong Kong.* Wikipedia. https://en.wikipedia.org/wiki/Hakka_people

Latour, B. (2005). *Reassembling the Social: An Introduction to Actor-Network-Theory*. Oxford University Press.

Lefebvre, H. (2004). *Rhythmanalysis: Space, Time and Everyday Life*. Continuum.

Leo, J. (2015). *Global Hakka: Hakka Identity in the Remaking*. Brill.

Lewicka, M. (2011). Place Attachment: How Far Have We Come in the Last 40 Years? *Journal of Environmental Psychology, 31*(3), 207–230.

Li, F. M. S. (2008). Culture as a Major Determinant in Tourism Development of China. *Current Issues in Tourism, 11*(6), 492–513. https://doi.org/10.1080/13683500802475786

Li, T. E., & McKercher, B. (2016). Effects of Place Attachment on Home Return Travel: A Spatial Perspective. *Tourism Geographies, 18*(4), 359–376. https://doi.org/10.1080/14616688.2016.1196238

Lin, S,Y. (1940, January). Salt Manufacture in Hong Kong. *The Hong Kong Naturalist*, Vol X(1), pp 14–20.

Lin, S. Y. (1967). Salt Manufacture in Hong Kong. *Journal of the Hong Kong Branch of the Royal Asiatic Society, 7*, 138–151.

Lin, Z. (林作尧). (2022, November 17). 客家人"根"在河洛，文化源远流长 (Hakka Culture). https://mp.weixin.qq.com/s/VpZjTa58RYu5DhGJFfhmSQ

Lisle, D. (2021). A Speculative Lexicon of Entanglement. *Millenium: Journal of International Studies, 49*(3), 435–461. https://doi.org/10.1177/03058298211021919

Low, S. M., & Altman, I. (Eds.). (2012). *Place Attachment: A Conceptual Inquiry*. Plenum Press.

Luehrmann, S. (2015). *Religion in Secular Archives: Soviet Atheism and Historical Knowledge*. Oxford University Press.

Lung, D. (2012). Built Heritage in Transition: A Critique of Hong Kong's Conservation Movement and the Antiquities and Monuments Ordinance. *Hong Kong Law Journal, 42*(1), 121–141.

Marcus, C. C. (2012). Environmental Memories. In S. M. Low & I. Altman (Eds.), *Place Attachment: A Conceptual Inquiry* (Ch.5, pp. 87–112). Plenum Press.

Marcinek, A. A., & Hunt, C. A. (2015). Social Capital, Ecotourism, and Empowerment in Shiripuno, Ecuador. *International Journal of Tourism Anthropology, 4*, 327 [Google Scholar].

Martin, B. D., & Schwab, E. (2012). Symbiosis: "Living Together" in Chaos. *Studies in the History of Biology, 4*(4), 7–25.

Maruyama, N. U., Weber, I., & Stronza, A. L. (2010). Negotiating Identity: Experiences of "Visiting Home" Among Chinese Americans. *Tourism, Culture & Communication, 10*(1), 1–14. https://doi.org/10.3727/109830410X12629765735551

Meyer, B., & de Witte, M. (2013). Heritage and the Sacred: Introduction. *Material Religion, 9*(3), 274–280. https://doi.org/10.2752/175183413X13730330868870

Mitchell, R. K., Agle, B. R., & Wood, D. J. (1997). Toward a Theory of Stakeholder Identification and Salience: Defining the Principle of Who and What Really Counts. *The Academy of Management Review, 22*(4), 853–886. https://doi.org/10.2307/259247

Mohajan, D., & Mohajan, H. (2022). *Constructivist Grounded Theory: A New Research Approach in Social Science* (MPRA Paper No. 114970), posted online 15 October 2022. https://mpra.ub.uni-muenchen.de/114970/

Müller, M. (2004). *Bibliography Concerning St. Arnold Janssen and St. Joseph Freinademetz*. Apud Collegium Verbi Divini.

Murphy, O. (2018, Spring). *Museum Studies as Critical Praxis: Developing an Active Approach to Teaching, Research and Practice* (Tate Papers, No. 29) [On-line research journal].

Murray, M. (2021). *The Camino de Santiago: Curating the Pilgrimage as Heritage and Tourism*. Berghahn Books.

Niedźwiedź, A., & Saraiva, C. (2015, June 22). *The Heritagization of Religious and Spiritual Practices: The Effects of Grassroots and Top-Down Policies*. SIEF Ethnology of Religion Working Group conference. Zagreb: International Society for Ethnology and Folklore. https://www.nomadit.co.uk/sief/sief2015/panels.php5?PanelID=3394

Olsen, D. H. (2003). Heritage, Tourism, and the Commodification of Religion. *Tourism Recreation Research, 28*(3), 99–104. https://doi.org/10.1080/02508281.2003.11081422

Olsen, D. H. (2019). Religion, Spirituality, and Pilgrimage in a Globalizing World. In D. J. Timothy (Ed.), *Handbook of Globalisation and Tourism* (pp. 270–283). Edward Elgar.

Olsen, D. H. (2022). Religious Tourism: A Spiritual or Touristic Experience? In R. Sharpley (Ed.), *Routledge Handbook of the Tourist Experience* (pp. 391–407). Routledge.

Olsen, D. H., & Timothy, D. J. (2021). *The Routledge Handbook of Religious and Spiritual Tourism*. Routledge.

Olson, B. D., Cooper, D. G., Viola, J. J., & Clark, B. (2016, 1 January). Community Narratives. In L. A. Jason & D. S. Glenwick (Eds.), *Handbook of Methodological Approaches to Community-Based Research: Qualitative, Quantitative, and Mixed Methods* (Online ed.). Oxford Academic. https://doi.org/10.1093/med:psych/9780190243654.001.0001

Oxford English Dictionary. (2020). Oxford University Press.

Oxford English Dictionary (2012). London: W.H. Smith

Parmar, B. L., Freeman, R. E., Harrison, J. S., Wicks, A. C., de Colle, S. & Purnell, L. (2010, June). *Stakeholder Theory: The State of the Art.* The Academy of Management Annals.

Peycam, P., Shu-Li, W., Yew-Foong, H., Hsin-Huang Michael, H. (Eds.). (2020). *Heritage as Aid and Diplomacy in Asia.* ISEAS—Yusof Ishak Institute; International Institute for Asian Studies (IIAS); and Institute of Sociology, Academia Sinica.

Phi, G., & Dredge, D. (2019). Critical issues in tourism co-creation. *Tourism Recreation Research, 44*(3), 281–283. https://doi.org/10.1080/02508281. 2019.1640492

Pimenova, K. (2022). Museums and Religious Heritage: Post-colonial and Post-socialist Perspectives. *Civilisations Revue Internationale d'anthropologie et de Sciences Humaines, 71,* 13–28.

Pletcher, K. (2013). *"Hakka".* Britannica. https://www.britannica.com/topic/Hakka

Pope John Paul II. (2003, October 5). Canonization of Three Blesseds. *Libreria Editrice Vaticana.* https://www.vatican.va/content/john-paul-ii/en/homilies/2003/documents/hf_jp-ii_hom_20031005_canonizations.html

Prahalad, C. K., & Ramaswamy, V. (2004). Co-creation Experiences: The Next Practice in Value Creation. *Journal of Interactive Marketing, 18*(3), 5–14. https://doi.org/10.1002/dir.20015

Preston, J. (1992). Spiritual Magnetism: An Organizing Principle for the Study of Pilgrimage. In A. Morinis (Ed.), *Sacred Journeys: The Anthropology of Pilgrimage* (pp. 31–46). Bloomsbury Academic Publishing.

Proshansky, H. M., Fabian, A. K., & Kaminoff, R. (1983). Place-Identity: Physical World Socialization of the Self. *Journal of Environmental Psychology, 3*(1), 57–83.

Richards, G., & Marques, L. (2012). Exploring Creative Tourism: Editors Introduction. *Journal of Tourism Consumption and Practice, 4*(2), 1–11.

Riley, R. B. (2012). Attachment to the Ordinary Landscape. In S. M. Low & I. Altman (Eds.), *Place Attachment: A Conceptual Inquiry* (ch.2, pp. 13–35). Plenum Press

Routledge, C., Wildschut, T., Sedikides, C., Juhl, J., & Arndt, J. (2012). The Power of the Past: Nostalgia as a Meaning-Making Resource. *Memory, 20*(5), 452–460.

Rushdie, S. (1991). *Imaginary Homelands—Essays and Criticism 1981-1991.* Granta in association with Penguin.

Sagarin, R. (2013, June 25). *To Overcome Your Company's Limits, Look to Symbiosis.* Harvard Business Review. https://hbr.org/2013/06/to-overcome-your-companys-limits-look-to

Said, E. W. (1984). *Orientalism* (2nd ed.). Vintage Press.

Salemink, O. (2009). Afterword: Questioning Faiths? Casting Doubts. In T. D. DuBois (Ed.), *Casting Faiths: Imperialism, Technology and the Transformation of Religion in East and Southeast Asia* (pp. 257–263). Palgrave Macmillan.

Salemink, O. (2016). Described, Inscribed, Written Off: Heritagisation as (Dis)Connection. In P. Taylor (Ed.), *Connected and Disconnected in Vietnam: Remaking Social Relations in a Post-socialist Nation* (pp. 311–346). Australian National University Press.

Salemink, O., Poulsen, R. R., & Ahl, S. I. (2020). Sacred But Not Holy: Awe, Spectacle, and the Heritage Gaze in Danish Religious Heritage Contexts. *Anthropological Notebooks, 26*(3), 70. https://doi.org/10.5281/zenodo.4604148

Santos, X. M. (2002). Pilgrimage and Tourism at Santiago De Compostela. *Tourism Recreation Research, 27*, 41–50.

Sarup, M., & Raja, T. (1996). *Identity, Culture, and the Postmodern World.* Edinburgh University Press.

Scannell, L., & Gifford, R. (2010). Defining Place Attachment: A Tripartite Organizing Framework. *Journal of Environmental Psychology, 30*(1), 1–10. https://doi.org/10.1016/j.jenvp.2009.09.006

Scheyvens, R. (1999). Ecotourism and the Empowerment of Local Communities. *Tourism Management, 20*, 245–249.

Scheyvens, R., & van der Watt, H. (2021). Tourism, Empowerment and Sustainable Development: A New Framework for Analysis. *Sustainability, 2021*(13), 12606. https://doi.org/10.3390/su132212606

Sedikides, C., & Wildschut, T. (2018). Finding Meaning in Nostalgia. *Review of General Psychology, 22*(1), 48–61. https://doi.org/10.1037/gpr0000109

Sham M. H., Chan C. S., Marafa, L. M., Cheung S. W., Shek K. F., & Lu Y. Y. (2022–On-going). *Community-Based Narratives and Public Experiential Engagement for Cultural and Historical Heritage Conservation and Revitalisation of Yim Tin Tsai, Sai Kung.* Chinese University of Hong Kong.

Shinde, K. A., & Olsen, D. H. (2023). Reframing the Intersections of Pilgrimage, Religious Tourism, and Sustainability. *Sustainability (Basel, Switzerland), 15*(1), 461–. https://doi.org/10.3390/su15010461

Singh, R. P. B. (2008). The Contestation of Heritage: The Enduring Importance of Religion. In B. Graham & P. Howard (Eds.), *Ashgate Research Companion to Heritage & Identity* (pp. 125–141). Ashgate Publishing.

Sinn, E. (1989). *Power and Charity: The Early History of the Tung Wah Hospital.* Oxford University Press.

Smith, L. (2006). *Uses of Heritage.* Routledge.

Smith, M. K. (2021). A New Spiritual Marketplace: Comparing New Age and New Religious Movements in an Age of Spiritual and Religious Tourism. In D. Olsen & D. Timothy (Eds.), *The Routledge Handbook of Religious and Spiritual Tourism* (Ch. 6). Routledge.

Sofield, T. H. B. (2001). Sustainability and Pilgrimage Tourism in the Kathmandu Valley of Nepal. In V. Smith & Brent, M. (Eds), *Hosts and Guests Re-visited: Tourism Issues of the 21st Century* (Ch. 20, pp. 257–274). Cognizant Communications Corp.

Sofield, T. H. B. (2003). *Empowerment for Sustainable Tourism Development.* Pergamon.

Sofield, T. H. B. (2013). Angkor: Tourism Management Caught in the Crossfire. In M. P. Oton, P. J. M. Mantinan, & V. P. Carril (Eds.), *Touristic Management of World Heritage Monuments and Cities* (pp. 117–152). Universidade de Santiago de Compostela Publications.

Sofield, T. H. B., Guia, J., & Specht, J. (2017, August). Organic 'Folkloric' Community Driven Place-Making and Tourism. *Journal of Tourism Management*, 1–22.

Sofield, T. H. B., & Li, F. M. S. (2020). Crown Cave, Guilin: A Chinese Perspective on Responsible Rural Tourism. In V. Nair, A. Hamzah, & G. Musa (Eds.), *Responsible Rural Tourism in Asia* (pp. 41–60). Channel View Publications.

Sofield, T. H. B., Li, F. M. S., Wong, G. H. Y., & Zhu, J. J. (2019). The Heritage of Chinese Cities as Seen Through the Gaze of Zhonghua Wenhua ('Chinese Common Knowledge'): Guilin as an Exemplar. *Journal of Heritage Tourism, Special Issue on 'Heritage and Cities in China'*, 12(3), 227–250.

Sofield, T. H. B., & Sivan, A. (1994). From Cultural Festival to International Sport—The Hong Kong Dragon Boat Races. *The Journal of Sport Tourism*, 1(3), 5–17. https://doi.org/10.1080/10295399408718541

Soja, E. W. (1996). *Thirdspace: Journeys to Los Angeles and Other Real-and-Imagined Places.* Blackwell.

Somerville, P. (1992). Homelessness and the Meaning of Home: Rooflessness or Rootlessness? *International Journal of Urban and Regional Research, 16*(4), 529–539. https://doi.org/10.1111/j.1468-2427.1992.tb00194.x

Steffen, P. (2012). Witness and Holiness, the Heart of the Life of Saint Joseph Freinademetz of Shandong. *Studia Missionalia, 61,* 257–392.

Suedtirolerland.it. (n.d.). *Birthplace of Saint Giuseppe Freinademetz—Pilgrimage Site.* https://www.suedtirolerland.it/en/highlights/sights/birthplace-of-saint-giuseppe-freinademetz/

Sullivan, B. M. (2015). *Sacred Objects in Secular Spaces Exhibiting Asian Religions in Museums.* Bloomsbury Academic, an imprint of Bloomsbury Publishing Plc.

Tanselle, G. T. (1998). *Literature and Artifacts.* Bibliographical Society of the University of Virginia.

The Art Newspaper. (2023, March 23). The 100 Most Popular Art Museums in the World—Who Has Recovered and Who Is Still Struggling? Visitor Figures 2022 Survey. *The Art Newspaper.* https://www.theartnewspaper.com/2023/03/27/the-100-most-popular-art-museums-in-the-worldwho-has-recovered-and-who-is-still-struggling

Thouki, A. (2022). Heritagization of Religious Sites: In Search of Visitor Agency and the Dialectics Underlying Heritage Planning Assemblages. *International Journal of Heritage Studies: IJHS, 28*(9), 1036–1065. https://doi.org/10. 1080/13527258.2022.2122535

Ticozzi, S. (2008). The Catholic Church and Nineteenth Century Village Life in Hong Kong. *Journal of the Hong Kong Branch of the Royal Asiatic Society, 48*, 111–149.

Timothy, D. J. (2011). *Cultural Heritage and Tourism: An Introduction.* Chanell View Publications.

Timothy, D. J. (Ed.). (2019). *Handbook of Globalisation and Tourism.* Edward Elgar Publishing.

Timothy, D. J. (2021). *Cultural Heritage and Tourism: An Introduction* (2nd ed.). Channel View Publications.

Torraco, R. J. (2016). Writing Integrative Literature Reviews: Using the Past and Present to Explore the Future. *Human Resource Development Review, 15*(4), 404–428. https://doi.org/10.1177/1534484316671606

Tsivolas, T. (2019). The Legal Foundations of Religious Cultural Heritage Protection. *Religions (Basel, Switzerland), 10*(4), 283–. https://doi.org/10. 3390/rel10040283

Tuan, Y.-F. (1974). *Topophilia: A Study of Environmental Perception, Attitudes and Values.* Prentice Hall.

Turner, B. (2006). Religion and Politics: Nationalism, Globalisation and Empire. *Asian Journal of Social Science, 34*(2), 209–224. https://doi.org/10.1163/ 156853106777371175

Turner, V. (1974a). Pilgrimage and Communitas. *Studia Missionalia, 23*, 305–307.

Turner, V. (1974b). *Dramas, Fields, and Metaphors: Symbolic Action in Human Society.* Cornell University Press.

UNESCO. (1994). *Nara Document on Authenticity.* UNESCO.

UNESCO. (2013). *Operational Guidelines for the Implementation of the World Heritage Convention.* UNESCO.

UNESCO. (2020). *Asia-Pacific Awards for Cultural Heritage Conservation, 2015–2019* (Vol. IV). UNESCO. https://unesdoc.unesco.org/ark:/48223/ pf0000374413.pdf. Accessed 22 November 2020.

UN-Habitat Commission. (2022). *Learn More About Us.* https://unhabi tat.org/about-us/learn-more#:~:text=UN%2DHabitat%20is%20the%20Unit ed,of%20adequate%20shelter%20for%20all

van der Tol, M., & Gorski, P. (2022). Secularisation as the Fragmentation of the Sacred and of Sacred Space. *Religion, State & Society, 50*(5), 495–512. https://doi.org/10.1080/09637494.2022.2144662

Van den Hemel, E., Salemink, O., and Stengs, I.L., (2022). Introduction: management of religion, sacralisation of heritage. In E. Van den Hemel, O.

Salemink, and I. L. Stengs, (Eds.), *Managing sacralities at religious heritage sites in contemporary Europe*. London: Berghahn Books, 1–21. Google Scholar

Van den Hemel, E., Salemink, O., and Stengs, I. (eds.) (2022). *Managing Sacralities: Competing and Converging Claims of Religious Heritage*. Berghahn Publications.

Venturini, T. (2010). Diving in Magma: How to Explore Controversies with Actor-Network Theory. *Public Understanding of Science, 19*(3), 258–273.

Vukonic, B. (2002). Religion, Tourism and Economics: A Convenient Symbiosis. *Tourism Recreation Research, 27*(2), 59–64. https://doi.org/10.1080/025 08281.2002.11081221

Wang, C., & Wong, S. L. (2014). Home as a Circular Process: A Study of the Indonesian Chinese in Hong Kong. In T. Mette (Ed.), *Beyond Chinatown: New Chinese Migrants and China's Global Expansion* (pp. 169–191). Nordic Institute of Asian Study Press.

Wang, S. L., Rowlands, M., & Zhu, Y. (Eds.). (2021). *Heritage and Religion in East Asia*. Routledge.

Watson, R. S. (1988). Remembering the Dead: Graves and Politics in Southeastern China. In J. Watson & E. Rawski (Eds.), *Death Ritual in Late Imperial and Modern China* (pp. 203–227). University of California Press.

Wildschut, T., Sedikides, C., Arndt, J., & Routledge, C. (2006). Nostalgia: Content, Triggers, Functions. *Journal of Personality and Social Psychology, 91*(5), 975–993.

Williams, M., & Webb, R. (1994). Rural Landscapes. In M. Williams (Ed.), *The Green Dragon: Hong Kong's Living Environment* (pp. 113–127). AbeBooks.

Wolf, A. P. (Ed.). (1974). *Religion and Ritual in Chinese Society*. Stanford University Press.

Wong, B. (1990). Chinese Customary Law—An Examination of T'sos and Family T'ongs. *Hong Kong Law Journal, 20*, 13–30.

Wong, T. S. (Dir.). (2013). *Then and Now—Yim Tin Tsai: A Hakka Catholic Village Reborn* [Documentary film]. New Wave Studios.

Xiao, X., Hui, Y.-F., & Peycam, P. M. F. (Eds.). (2017). *Citizens, Civil Society and Heritage-Making in Asia*. ISEAS Publishing.

Xu, H., & Sofield, T. H. B. (2017). New Interests of Urban Heritage and Tourism Research in Chinese Cities. *Journal of Heritage Tourism, 12*(3), 223–226. https://doi.org/10.1080/1743873X.2016.1244539

Xu, H., & Sofield, T. H. B. (Eds.). (2018). *Heritage Tourism and Cities in China*. Routledge.

Yanata, K., & Sharpley, R. (2021). Coexistence Between Tourists and Monks. Managing Temple-Stay Tourism at Koyasan, Japan. In D. H. Olsen & D. Timothy (Eds.), *The Routledge Handbook of Religious and Spiritual Tourism* (pp. 398–410). Routledge.

Yau, K. L. (2016). *From Invisible to Visible: Representations and Self-Representaions of Hakka Women In Hong Kong, 1900s–Present* (Master's thesis, Lingnan University, Hong Kong). https://commons.ln.edu.hk/cgi/viewcontent.cgi?article=1008&context=his_etd. Accessed 22 November 2023.

Zhang, ChaoZhi, Fyall, A. and Zheng, Yenfen (2010). Conflict within World Heritage Sites in China: A Longitudinal Study. *Current Issues In Tourism, 18*(2): 110–116 https://doi.org/10.1080/13683500.2014.912204.

Zhang, Y., & Wu, Z. (2016). The Reproduction of Heritage in a Chinese Village: Whose Heritage, Whose Pasts? *International Journal of Heritage Studies, 22*(3), 228–241.

Zheng, S. (2023). The Heritagisation of Rituals: Commodification and Transmission. A Case Study of Nianli Festival in Zhanjiang, China. *Études Mongoles & Sibériennes, Centrasiatiques & Tibétaines, 54.* https://journals.openedition.org/emscat/6109?lang=en

INDEX

GPSR Compliance

The European Union's (EU) General Product Safety Regulation (GPSR) is a set of rules that requires consumer products to be safe and our obligations to ensure this.

If you have any concerns about our products, you can contact us on ProductSafety@springernature.com

In case Publisher is established outside the EU, the EU authorized representative is:

Springer Nature Customer Service Center GmbH
Europaplatz 3
69115 Heidelberg, Germany

The manufacturer's authorised representative in the EU is Springer Nature Customer Service Centre GmbH, Europaplatz 3, 69115 Heidelberg, Germany. If you have any concerns regarding our products, please contact ProductSafety@springernature.com

Printed and bound by CPI Group (UK) Ltd, Croydon, CR0 4YY

29/04/2026

02099552-0002